HARCOURT

Math

Reteach
Workbook

TEACHER'S EDITION
Grade 5

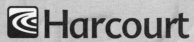

Harcourt

Orlando • Boston • Dallas • Chicago • San Diego
www.harcourtschool.com

© Harcourt

CONTENTS

Understand Place Value

The number 391,568 may be easier to read and write if you use a place-value chart.

Thousands				Ones		
Hundreds	Tens	Ones		Hundreds	Tens	Ones
3	0	0	,	0	0	0
	9	0	,	0	0	0
		1	,	0	0	0
				5	0	0
					6	0
						8

Standard form: 391,568
Expanded form: 300,000 + 90,000 + 1,000 + 500 + 60 + 8
Word form: Three hundred ninety-one thousand,
five hundred sixty-eight

Write the number in the place-value chart. Then write the number in expanded form.

1. 716,583

Thousands				Ones		
H	T	O		H	T	O
7	0	0	,	0	0	0
	1	0	,	0	0	0
		6	,	0	0	0
				5	0	0
					8	0
						3

700,000 + 10,000 + 6,000 +

500 + 80 + 3

2. 78,056

Thousands				Ones		
H	T	O		H	T	O
	7	0	,	0	0	0
		8	,	0	0	0
				0	0	0
					5	0
						6

70,000 + 8,000 +

50 + 6

Use the place-value chart to help you write the value of the **bold-faced** digit.

3. 5**8**,346

8,000

4. 7**2**3,308

20,000

5. **4**68,005

400,000

6. 420,**8**22

800

Millions and Billions

You can use a place-value chart to help you read and write greater numbers such as 721,306,984.

Millions			Thousands			Ones			
H	**T**	**O**	**H**	**T**	**O**	**H**	**T**	**O**	
7	0	0	0	0	0	0	0	0	seven hundred
	2	0	0	0	0	0	0	0	twenty-one million,
		1	0	0	0	0	0	0	
			3	0	0	0	0	0	three hundred six
				0	0	0	0	0	thousand,
					6	0	0	0	
						9	0	0	nine hundred
							8	0	eighty-four
								4	

Standard form: 721,306,984
Expanded form: 700,000,000 + 20,000,000 + 1,000,000 + 300,000 + 6,000 + 900 + 80 + 4
Word form: seven hundred twenty-one million, three hundred six thousand, nine hundred eighty-four

---

Write the number in word form.

1. 2,267,025,142

two billion, two hundred

sixty-seven million,

twenty-five thousand,

one hundred forty-two

2. 702,326,500

seven hundred two million,

three hundred twenty-six

thousand, five hundred

Write the number in standard form.

3. 600,000,000 + 50,000,000 + 9,000,000 + 800,000 + 40,000 + 3,000 + 700 + 1

659,843,701

4. thirty-five billion, eight hundred six million, four hundred eighty-six thousand, two hundred twenty-six

35,806,486,226

Compare Numbers

You can use a place-value chart to compare numbers.

Thousands			Ones		
H	T	O	H	T	O
2	8	9	8	6	5
2	8	9	7	6	5

Compare the digits from left to right.

← First number: 289,865

← Second number: 289,765

↑   ↑   ↑   ↑
same   same   same   8 > 7

So, 289,865 > 289,765.

Complete the place-value chart. Write <, >, or = for each ◯.

1.

Thousands			Ones		
H	T	O	H	T	O
3	7	5	8	4	1
3	6	7	8	4	1

375,841 ⟩ 367,841

2.

Thousands			Ones		
H	T	O	H	T	O
6	7	7	8	6	0
6	7	7	8	6	0

677,860 ⟩ 677,860

3.

Thousands			Ones		
H	T	O	H	T	O
4	6	7	9	3	5
4	7	6	9	3	5

467,935 ⟨ 476,935

4.

Thousands			Ones		
H	T	O	H	T	O
9	8	6	4	9	6
9	8	6	4	9	5

986,496 ⟩ 986,495

5.

Millions			Thousands			Ones		
H	T	O	H	T	O	H	T	O
	4	7	2	0	6	3	8	5
	4	7	0	8	3	2	1	9

47,206,385 ⟩ 47,083,219

Order Numbers

You can use a place-value chart to order numbers. Compare the digits from left to right.

Thousands			Ones		
H	T	O	H	T	O
3	2	2	6	7	8
3	4	2	1	9	8
3	2	2	5	0	1

↑   ↑   ↑   ↑
same 4 > 2 same 6 > 5

So, 342,198 > 322,678 > 322,501.

Since 4 > 2, 342,198 is the greatest number.

Continue to compare with the remaining two numbers.

Since 6 > 5, 322,678 > 322,501.

Use the place-value chart to order the numbers.

1. 144,421; 144,321; 145,221

Thousands			Ones		
H	T	O	H	T	O
1	4	4	4	2	1
1	4	4	3	2	1
1	4	5	2	2	1

145,221 > 144,421 > 144,321

2. 532,124; 58,124; 532,876

Thousands			Ones		
H	T	O	H	T	O
5	3	2	1	2	4
	5	8	1	2	4
5	3	2	8	7	6

532,876 > 532,124 > 58,124

3. 456,342,523; 456,342,876; 494,123,563

Millions			Thousands			Ones		
H	T	O	H	T	O	H	T	O
4	5	6	3	4	2	5	2	3
4	5	6	3	4	2	8	7	6
4	9	4	1	2	3	5	6	3

494,123,563 > 456,342,876 > 456,342,523

Name _____

Problem Solving Skill

Use a Table

Tables help organize data so you can make comparisons.

Suppose you want to compare the sizes of four planets.
You could make the following table.

THE PLANETS	
Name	Diameter (miles)
Mercury	3,030
Venus	7,517
Earth	7,921
Mars	4,222

Mercury Venus Earth Mars

• Look at the diameters. Compare the digits from left to right.

• The smallest planet is Mercury. The largest planet is Earth.

Use the tables to answer the questions.

1. This table shows the sales for a popular music store chain. Which type of music had the greatest sales amount? the least sales amount?

_____Country_____

_____Light Rock_____

MUSIC SALES	
Music Type	Sales (in dollars)
Alternative Rock	1,345,850
Classical	548,290
Country	1,930,000
Light Rock	425,830

2. This table shows the areas of some of the world's oceans. Which of these oceans has the greatest area? the least area?

North Pacific; South Atlantic

_____

OCEANS	
Name	Area (square miles)
Indian	31,507,000
North Pacific	32,225,000
South Pacific	25,298,000
North Atlantic	18,059,000
South Atlantic	14,426,000

© Harcourt

Tenths and Hundredths

Money can be used to model decimals.

A dollar represents one whole, or $1.00.

one
1.0
1

The whole is divided into 10 equal parts. One dime is $\frac{1}{10}$ of a dollar, or $0.10.

one tenth
0.1
$\frac{1}{10}$

The whole is divided into 100 equal parts. One penny is $\frac{1}{100}$ of a dollar, or $0.01.

one hundredth
0.01
$\frac{1}{100}$

Write as a decimal.

1.

0.3

2.

0.47

3.

2.05

4. 1 dollar, 2 dimes, and 9 pennies

1.29

5. 3 dollars and 6 dimes

3.6

6. 7 dollars, 5 dimes, and 7 pennies

7.57

Write as a decimal and a fraction.

7. 4 dimes and 6 pennies

$0.46, \frac{46}{100}$

8. 2 dollars and 7 dimes

$2.7, 2\frac{7}{10}$

9. 3 dollars, 6 dimes, and 5 pennies

$3.65, 3\frac{65}{100}$

10. five and six tenths

$5.6, 5\frac{6}{10}$

11. five hundredths

$0.05, \frac{5}{100}$

12. four and three tenths

$4.3, 4\frac{3}{10}$

13. one and eighty-three hundredths

$1.83, 1\frac{83}{100}$

14. nine and seventeen hundredths

$9.17, 9\frac{17}{100}$

15. two and eight tenths

$2.8, 2\frac{8}{10}$

Thousandths and Ten-Thousandths

A place-value chart can help you find the value of each digit in a decimal.

	Ones	Tenths	Hundredths	Thousandths	Ten-Thousandths
Decimal:	2 •	3	6	5	1
Read:	two	three tenths	six hundredths	five thousandths	one ten-thousandth
Write:	2.0	0.3	0.06	0.005	0.0001

In *Standard Form:* 2.3651
In *Expanded Form:* 2.0 + 0.3 + 0.06 + 0.005 + 0.0001
In *Word Form:* two and three thousand, six hundred fifty-one ten-thousandths

---

Record each decimal in the place-value chart. Write each decimal in expanded form and word form.

1. 1.5138

Ones	Tenths	Hundredths	Thousandths	Ten-Thousandths
1 •	5	1	3	8

1.0 + 0.5 + 0.01 + 0.003 + 0.0008;

one and five thousand, one hundred thirty-eight ten-thousandths

2. 4.973

Ones	Tenths	Hundredths	Thousandths	Ten-Thousandths
4 •	9	7	3	

4.0 + 0.9 + 0.07 + 0.003;

four and nine hundred seventy-three thousandths

3. 7.0458

Ones	Tenths	Hundredths	Thousandths	Ten-Thousandths
7 •	0	4	5	8

7.0 + 0.04 + 0.005 + 0.0008;

seven and four hundred fifty-eight ten-thousandths

Equivalent Decimals

Equivalent decimals name the same
number or amount.

2 tenths = 20 hundredths

0.2 = 0.20

In the place-value chart, both numbers have a 2 in the tenths place.

Ones	Tenths	Hundredths
0	2	
0	2	0

← 2 tenths
← 20 hundredths

The zero to the right of the 2 does not change the value of
the decimal. So, they are equivalent.

Write the decimals in the place-value chart. Then write *equivalent* or *not equivalent* to describe each pair.

1. 2.5 and 2.50

Ones	Tenths	Hundredths
2	5	
2	5	0

_____ equivalent _____

2. 0.73 and 0.703

Ones	Tenths	Hundredths	Thousandths
0	7	3	
0	7	0	3

_____ not equivalent _____

Write the two decimals that are equivalent.

3. 3.05
3.050
3.500

3.05 and 3.050

4. 1.110
1.1
1.11

1.110 and 1.11

5. 0.180
0.0180
0.018

0.0180 and
0.018

6. 7.77
7.707
7.770

7.77 and 7.770

Write an equivalent decimal for each number. Possible answers are given.

7. 0.05 _____ 0.050 _____

8. 2.100 _____ 2.1 _____

9. 2.875 _____ 2.8750 _____

10. 0.040 _____ 0.04 _____

Compare and Order Decimals

You can use a place-value chart to compare 6.741 and 6.742.

Ones	Tenths	Hundredths	Thousandths
6	7	4	1
6	7	4	2

↑   ↑   ↑   ↑

Equal   Equal   Equal   2 > 1
numbers   numbers   numbers of
of ones   of tenths   hundredths

So, 6.742 > 6.741.

Write the numbers in the place-value chart. Then write <, >, or = in each ◯.

1. 2.45 ◯< 2.54

O.	T	H
2	4	5
2	5	4

2. 6.23 ◯= 6.230

O.	T	H	Th
6	2	3	
6	2	3	0

3. 72.648 ◯< 72.658

T	O.	T	H	Th
7	2	6	4	8
7	2	6	5	8

4. 564.876 ◯> 564.786

H	T	O.	T	H	Th
5	6	4	8	7	6
5	6	4	7	8	6

Write <, >, or = in each ◯.

5. 3.21 ◯= 3.210

6. 721.460 ◯> 72.146

7. 6.275 ◯> 6.257

8. 468.036 ◯< 468.136

Order from least to greatest.

9. 16.54, 16.56, 16.55 _____ 16.54 < 16.55 < 16.56

10. 3.400, 3.004, 3.040 _____ 3.004 < 3.040 < 3.400

Problem Solving Skill

Draw Conclusions

Michael exercises at 4:00 P.M. daily unless he is sick. The table shows the number of hours Michael exercised last week.

Can the conclusion be drawn from the information given? Write *yes* or *no*. Explain your choice.

HOURS MICHAEL EXERCISED	
Day	Hours
Monday	1.8
Tuesday	1.5
Wednesday	0
Thursday	2.2
Friday	1.6

Michael usually eats dinner at 5:30.

No; from 4:00 to 5:30 is

1.5 hours. On Monday, Thursday,

and Friday he was still

exercising, not eating.

Michael was sick on Wednesday.

Yes; he exercises unless

he is sick. Since he didn't

exercise on Wednesday

he was probably sick.

Can the conclusion be drawn from the information given? Write *yes* or *no*. Explain your choice.

During a kickball game between two gym classes, the final score was 25 to 22. Each team had 15 players.

1. There were more boys on the winning team than on the losing team.

No; there is no information about

the numbers of boys and girls

on the two teams.

2. There was a winning team.

Yes; the team that scored

25 points won because

25 > 22.

3. Each player kicked a homerun.

No; it doesn't tell if any players

kicked homeruns.

4. More than 40 points were scored in the game.

Yes; there was a total of 47

points scored in the game.

Round Whole Numbers

You can round whole numbers by using the rounding rules.

Step 1: UNDERLINE the digit in the place to which you want to round.

Step 2: COMPARE the digit to the right of the underlined digit to 5.
Round Down: If the digit to the right is less than 5, the underlined digit stays the same.
Round Up: If the digit to the right is 5 or greater, increase the underlined digit by 1.

Step 3: CHANGE all digits to the right of the underlined digit to zeros.

A. Round 43,658 to the nearest hundred.
UNDERLINE. 43,658
COMPARE. 5 = 5
Round Up.
CHANGE. 43,700

B. Round 9,309,587 to the nearest million.
UNDERLINE. 9,309,587
COMPARE. 3 < 5
Round Down.
CHANGE. 9,000,000

Round each number to the place of the **bold-faced** digit.

1. 38,**7**61

UNDERLINE 38,761

COMPARE __7__ ⓧ 5

Round __Up__

CHANGE __39,000__

2. 719,**4**32

UNDERLINE 719,432

COMPARE __2__ ⓧ 5

Round __Down__

CHANGE __719,430__

Round 2,409,485 to the place named.

3. hundred thousands

UNDERLINE 2,409,485

COMPARE __0__ ⓧ 5

Round __Down__

CHANGE __2,400,000__

4. hundreds

UNDERLINE 2,409,485

COMPARE __8__ ⓧ 5

Round __Up__

CHANGE __2,409,500__

Estimate Sums and Differences

You can estimate sums and differences by rounding the numbers in the problem before performing the operation. One way to round is to round to the greatest place-value position. For example:

A. Estimate the **sum** by rounding.

$$
\begin{array}{rcr}
3,709,525 & \rightarrow & 4,000,000 \\
+\ \ 567,802 & \rightarrow & +\ \ 600,000 \\
\hline
& & 4,600,000
\end{array}
$$

The sum is about 4,600,000.

B. Estimate the **difference** by rounding.

$$
\begin{array}{rcr}
539,014 & \rightarrow & 500,000 \\
-205,918 & \rightarrow & -200,000 \\
\hline
& & 300,000
\end{array}
$$

The difference is about 300,000.

Estimate by rounding. Possible estimates are given.

1.
$$
\begin{array}{rcr}
473,542 & \rightarrow & 500,000 \\
+207,958 & \rightarrow & +200,000 \\
\hline
& & 700,000
\end{array}
$$

2.
$$
\begin{array}{rcr}
741,356 & \rightarrow & 700,000 \\
-157,900 & \rightarrow & -200,000 \\
\hline
& & 500,000
\end{array}
$$

3.
$$
\begin{array}{rcr}
8,619,724 & \rightarrow & 9,000,000 \\
+3,970,685 & \rightarrow & +4,000,000 \\
\hline
& & 13,000,000
\end{array}
$$

4.
$$
\begin{array}{rcr}
5,101,118 & \rightarrow & 5,000,000 \\
-\ \ 496,007 & \rightarrow & -\ \ 500,000 \\
\hline
& & 4,500,000
\end{array}
$$

5.
$$
\begin{array}{rcr}
724,581 & \rightarrow & 700,000 \\
-219,067 & \rightarrow & -200,000 \\
\hline
& & 500,000
\end{array}
$$

6.
$$
\begin{array}{rcr}
192,837 & \rightarrow & 200,000 \\
+445,672 & \rightarrow & +400,000 \\
\hline
& & 600,000
\end{array}
$$

7.
$$
\begin{array}{rcr}
521,739 & \rightarrow & 500,000 \\
+659,931 & \rightarrow & +700,000 \\
\hline
& & 1,200,000
\end{array}
$$

8.
$$
\begin{array}{rcr}
911,011 & \rightarrow & 900,000 \\
+187,408 & \rightarrow & +200,000 \\
\hline
& & 1,100,000
\end{array}
$$

9.
$$
\begin{array}{rcr}
4,516,361 & \rightarrow & 5,000,000 \\
+3,497,205 & \rightarrow & +3,000,000 \\
\hline
& & 8,000,000
\end{array}
$$

10.
$$
\begin{array}{rcr}
6,212,345 & \rightarrow & 6,000,000 \\
-3,493,968 & \rightarrow & -3,000,000 \\
\hline
& & 3,000,000
\end{array}
$$

Name _____

Add and Subtract Whole Numbers

You can add or subtract to find an exact answer.

Estimates will help you determine if you have a reasonable answer.

Suppose you have saved 3,857 pennies. Then your mom gives you 2,234 more pennies to help you buy a present for a friend. How many pennies do you have altogether?

First, estimate.

$$
\begin{array}{r}
3,857 \\
+2,234 \\
\end{array}
\quad
\begin{array}{c}
\rightarrow \\
\rightarrow
\end{array}
\quad
\begin{array}{r}
4,000 \\
+2,000 \\
\hline
6,000
\end{array}
$$

The answer should be close to 6,000.

Then, add to find the exact answer.

$$
\begin{array}{r}
\overset{1}{3}\,\overset{}{8}\,\overset{1}{5}\,7 \\
+2\,2\,3\,4 \\
\hline
6\,0\,9\,1
\end{array}
$$

6,091 is close to the estimate, so the answer is reasonable. You have 6,091 pennies.

Estimate. Then find the exact sum or difference. Possible estimates are given.

1.
$$
\begin{array}{r}
5\,4\,9\,2 \\
+4\,0\,7\,8 \\
\hline
9\,5\,7\,0
\end{array}
\quad
\begin{array}{r}
\rightarrow \\
\rightarrow
\end{array}
\quad
\begin{array}{r}
5,000 \\
+4,000 \\
\hline
9,000
\end{array}
$$

2.
$$
\begin{array}{r}
7\,9\,0\,6 \\
-4\,2\,3\,4 \\
\hline
3\,6\,7\,2
\end{array}
\quad
\begin{array}{r}
\rightarrow \\
\rightarrow
\end{array}
\quad
\begin{array}{r}
8,000 \\
-4,000 \\
\hline
4,000
\end{array}
$$

3.
$$
\begin{array}{r}
2\,9\,5\,3\,6 \\
-1\,0\,8\,1\,9 \\
\hline
1\,8\,7\,1\,7
\end{array}
\quad
\begin{array}{r}
\rightarrow \\
\rightarrow
\end{array}
\quad
\begin{array}{r}
30,000 \\
-10,000 \\
\hline
20,000
\end{array}
$$

4.
$$
\begin{array}{r}
6\,8\,4\,4 \\
+4\,7\,3\,9 \\
\hline
11\,5\,8\,3
\end{array}
\quad
\begin{array}{r}
\rightarrow \\
\rightarrow
\end{array}
\quad
\begin{array}{r}
7,000 \\
+5,000 \\
\hline
12,000
\end{array}
$$

5.
$$
\begin{array}{r}
1\,3\,7\,6 \\
-\ \ 4\,3\,2 \\
\hline
9\,4\,4
\end{array}
\quad
\begin{array}{r}
\rightarrow \\
\rightarrow
\end{array}
\quad
\begin{array}{r}
1,400 \\
-\ \ 400 \\
\hline
1,000
\end{array}
$$

6.
$$
\begin{array}{r}
3\,6\,7\,4\,8 \\
+1\,4\,2\,4\,7 \\
\hline
5\,0\,9\,9\,5
\end{array}
\quad
\begin{array}{r}
\rightarrow \\
\rightarrow
\end{array}
\quad
\begin{array}{r}
40,000 \\
+10,000 \\
\hline
50,000
\end{array}
$$

Choose a Method

You add and subtract greater numbers the same way you add and subtract smaller numbers.

It may become difficult to keep place values aligned when adding and subtracting greater numbers. Commas help you to line up the numbers.

For example, find the sum of 6,716,678 and 5,014,209.

- Line up the addends along the commas.

- Add to find the exact answer.

$$\begin{array}{r} \overset{1}{}\overset{1}{} \\ 6,716,678 \\ +\ 5,014,209 \\ \hline 11,730,887 \end{array} \quad \rightarrow \quad \begin{array}{r} 7,000,000 \\ +\ 5,000,000 \\ \hline 12,000,000 \end{array}$$

- Estimate the sum to see if your answer is reasonable.

11,730,887 is close to the estimate of 12,000,000, so the answer is reasonable.

Find the sum or difference. Estimate to check. Possible estimates are given.

1.
$$\begin{array}{r} 8,432,790 \\ +\ 3,876,339 \\ \hline 12,309,129 \end{array} \quad \begin{array}{l} \rightarrow \\ \rightarrow \end{array} \quad \begin{array}{r} 8,000,000 \\ +\ 4,000,000 \\ \hline 12,000,000 \end{array}$$

2.
$$\begin{array}{r} 4,918,471 \\ -\ 1,839,220 \\ \hline 3,079,251 \end{array} \quad \begin{array}{l} \rightarrow \\ \rightarrow \end{array} \quad \begin{array}{r} 5,000,000 \\ -\ 2,000,000 \\ \hline 3,000,000 \end{array}$$

3.
$$\begin{array}{r} 9,010,776 \\ -\ 4,573,932 \\ \hline 4,436,844 \end{array} \quad \begin{array}{l} \rightarrow \\ \rightarrow \end{array} \quad \begin{array}{r} 9,000,000 \\ -\ 5,000,000 \\ \hline 4,000,000 \end{array}$$

4.
$$\begin{array}{r} 3,825,449 \\ +\ 4,361,749 \\ \hline 8,187,198 \end{array} \quad \begin{array}{l} \rightarrow \\ \rightarrow \end{array} \quad \begin{array}{r} 4,000,000 \\ +\ 4,000,000 \\ \hline 8,000,000 \end{array}$$

Copy the problem. Use commas to help you line up the numbers. Find the sum or difference. Estimate to check. Possible estimates are given.

5. 6,654,148 + 4,732,387

$$\begin{array}{r} 6,654,148 \\ +\ 4,732,387 \\ \hline 11,386,535 \end{array} \quad \begin{array}{l} \rightarrow \\ \rightarrow \end{array} \quad \begin{array}{r} 7,000,000 \\ +\ 5,000,000 \\ \hline 12,000,000 \end{array}$$

6. 7,927,881 − 4,618,532

$$\begin{array}{r} 7,927,881 \\ -\ 4,618,532 \\ \hline 3,309,349 \end{array} \quad \begin{array}{l} \rightarrow \\ \rightarrow \end{array} \quad \begin{array}{r} 8,000,000 \\ -\ 5,000,000 \\ \hline 3,000,000 \end{array}$$

Name _____

Problem Solving Strategy

Use Logical Reasoning

A table can help you with logical reasoning.

Elizabeth, Alan, Calvin, and Marie each ordered a different ice cream flavor. The flavor choices were vanilla, peach, chocolate, and strawberry. Neither Alan nor Marie ordered vanilla. Calvin had a brown ice cream stain on his t-shirt. Marie is allergic to strawberries. Which flavor ice cream did each person order?

- Calvin had a brown stain on his t-shirt. Put a *yes* in the chocolate column for Calvin and a *no* in each empty box in that row and in that column.

- Marie is allergic to strawberries and she did not order vanilla. Put a *no* in those boxes. Put a *yes* in the remaining box, peach, and a *no* in the remaining boxes in that column.

	vanilla	peach	chocolate	strawberry
Elizabeth	Yes	No	No	No
Alan	No	No	No	Yes
Calvin	No	No	Yes	No
Marie	No	Yes	No	No

- Alan did not order vanilla. Put a *no* in that box. That leaves strawberry.

- So, Elizabeth ordered vanilla. Put a *yes* in that box.

Use logical reasoning and the table to solve.

1. Rishawn, Julie, Kevin, and LaTia each have a different favorite subject. Julie likes to use paint and chalk. LaTia enjoys using numbers. Science is not Kevin's favorite subject. What is each student's favorite subject?

	art	math	music	science
Rishawn	No	No	No	Yes
Julie	Yes	No	No	No
Kevin	No	No	Yes	No
LaTia	No	Yes	No	No

Rishawn, science; Julie, art; Kevin, music; LaTia, math

Round Decimals

The same rules you learned for rounding whole numbers can be used to round decimals.

Step 1: <u>Underline</u> the digit in the place to which you want to round.

Step 2: Compare the digit at the right of the underlined digit to 5.
Round Down: If the digit at the right is less than 5, the underlined digit stays the same.
Round Up: If the digit at the right is 5 or more, increase the underlined digit by 1.

Step 3: Rewrite all digits to the right of the underlined digit as zeros. An equivalent decimal can be written by leaving off trailing zeros.

A. Round 5.6431 to the nearest hundredth.	**B.** Round 0.8287 to the nearest thousandth.
Underline. 5.6<u>4</u>31	Underline. 0.82<u>8</u>7
Compare. 3 < 5 Round down.	Compare. 7 > 5 Round up.
Rewrite. 5.6400 or 5.64	Rewrite. 0.8290 or 0.829

---

1. Round 4.**1**872 to the place of the **bold-faced** digit.

Underline.   4.**1**872

Compare.   <u>8</u> ⊘ 5   Round <u>up</u>.

Rewrite. <u>   4.2000 or 4.2   </u>

2. Round 82.64751 to the nearest thousandth.

Underline.   82.6475

Compare.   <u>5</u> ⊖ 5   Round <u>up</u>.

Rewrite. <u> 82.64800 or 82.648 </u>

Round each number to the place of the **bold-faced** digit.

3. 7.**3**25

<u>   7.3   </u>

4. 9.0**2**87

<u>   9.03   </u>

5. 108.1**0**8

<u>   108.11   </u>

6. 2**6**.3199

<u>   26   </u>

Round 12.8405 to the place named.

7. hundredths

<u>   12.84   </u>

8. ones

<u>   13   </u>

9. tenths

<u>   12.8   </u>

10. thousandths

<u>   12.841   </u>

Estimate Decimal Sums and Differences

Jonas earned $25.87. Kevin earned $20.94. About how much did they earn in all? About how much more did Jonas earn than Kevin?

You can estimate decimal sums and differences by rounding the amounts to the nearest whole number and then adding or subtracting.

A. Estimate the sum by rounding.	**B.** Estimate the difference by rounding.
$\begin{array}{rcr} \$25.87 & \to & \$26 \\ +\ 20.94 & \to & +\ 21 \\ \hline & & \$47 \end{array}$	$\begin{array}{rcr} \$25.87 & \to & \$26 \\ -\ 20.94 & \to & -\ 21 \\ \hline & & \$5 \end{array}$
They earned about $47.	Jonas earned about $5 more than Kevin.

Estimate the sum or difference by rounding to the nearest whole number or dollar.

1. $\begin{array}{rcr} \$63.98 & \to & \$64 \\ +\ 5.29 & \to & +\ 5 \\ \hline & & \$69 \end{array}$

2. $\begin{array}{rcr} 9.684 & \to & 10 \\ -2.395 & \to & -\ 2 \\ \hline & & 8 \end{array}$

3. $\begin{array}{rcr} 25.39 & \to & 25 \\ -17.71 & \to & -\ 18 \\ \hline & & 7 \end{array}$

Estimate the sum or difference to the nearest tenth.

4. $\begin{array}{rcr} 8.604 & \to & 8.6 \\ -\ 6.71 & \to & -6.7 \\ \hline & & 1.9 \end{array}$

5. $\begin{array}{rcr} 26.4572 & \to & 26.5 \\ +11.3518 & \to & +11.4 \\ \hline & & 37.9 \end{array}$

6. $\begin{array}{rcr} 56.8 & \to & 56.8 \\ +\ 8.592 & \to & +\ 8.6 \\ \hline & & 65.4 \end{array}$

Estimate the sum or difference. Possible estimates are given. Methods may vary.

7. $\begin{array}{rcr} 8.453 & \to & 8 \\ -\ 1.21 & \to & -1 \\ \hline & & 7 \end{array}$

8. $\begin{array}{rcr} 8.25 & \to & 8 \\ +0.385 & \to & +0 \\ \hline & & 8 \end{array}$

9. $\begin{array}{rcr} 9.52 & \to & 10 \\ +\ 1.29 & \to & +\ 1 \\ \hline & & 11 \end{array}$

10. $\begin{array}{rcr} 7.05 & \to & 7 \\ -\ 0.63 & \to & -1 \\ \hline & & 6 \end{array}$

11. $\begin{array}{rcr} 5.128 & \to & 5 \\ -\ 1.56 & \to & -2 \\ \hline & & 3 \end{array}$

12. $\begin{array}{rcr} 2.31 & \to & 2 \\ +4.804 & \to & +5 \\ \hline & & 7 \end{array}$

Add and Subtract Decimals

To add or subtract decimals, line up the decimal points in the problem. Finding an estimate first will help you determine if your answer is reasonable.

First, estimate.	**Then, subtract to find the exact answer.**
18.948 → 19 − 5.765 → − 6 ———— 13	

The answer should be close to 13.

13.183 is close to the estimate, so the answer is reasonable.

The exact subtraction:

	1	8	•	⁸ø̸	¹⁴4̸	8
−		5	•	7	6	5
	1	3	•	1	8	3

Estimate. Then find the exact sum or difference. Possible estimates are given.

1.
```
    1 • 5 8 →   2
 + 4 • 5 3 → + 5
   6 • 1 1     7
```

2.
```
   1 8 • 5 2 →  19
 +   3 • 7 3 → + 4
   2 2 • 2 5    23
```

3.
```
    6 • 3 9 →   6
    2 • 1 8 →   2
 + 7 • 8 5 → + 8
  16 • 4 2     16
```

4.
```
   8 • 7 6 →   9
 − 5 • 2 3 → − 5
   3 • 5 3     4
```

5.
```
   1 6 • 3 2 →  16
 −   4 • 8   → − 5
   1 1 • 5 2    11
```

6.
```
   6 • 2 8 →   6
 − 3 • 9 6 → − 4
   2 • 3 2     2
```

7. $5.86 + 8.79 = n$

$n = 14.65$

8. $14.09 - 2.87 = n$

$n = 11.22$

Zeros in Subtraction

Find $1.34 - 1.256$.

- To subtract decimal numbers, line up the numbers along the decimal points.

$$\begin{array}{r} 1.34 \\ -\,1.256 \end{array}$$

- Add zeros so both numbers have the same number of decimal places.

$$\begin{array}{r} 1.34\mathbf{0} \\ -\,1.256 \end{array}$$

- Subtract.

- Place a decimal point in the answer, below the decimal points in the problem.

$$\begin{array}{r} \overset{2\ 13\ 10}{1.3\cancel{4}\cancel{0}} \\ -\,1.256 \\ \hline 0.084 \end{array}$$

So, $1.34 - 1.256 = 0.084$.

Find the difference.

| 1. | 2.7
 − 1.5
 1.2 | 2. | 3.94
 − 2.6
 1.34 | 3. | 4.75
 − 2.56
 2.19 | 4. | 6.8
 − 3.9
 2.9 | 5. | 5.1
 − 3.08
 2.02 |

| 6. | 3.5
 − 2.8
 0.7 | 7. | 4.4
 − 1.65
 2.75 | 8. | 7.643
 − 3.4
 4.243 | 9. | 11.904
 − 8.626
 3.278 | 10. | 16.24
 − 9.1
 7.14 |

| 11. | 4.2
 − 2.83
 1.37 | 12. | 5.6
 − 3.58
 2.02 | 13. | 9.41
 − 6.527
 2.883 | 14. | 14.5
 − 8.872
 5.628 | 15. | 35.4
 − 15.567
 19.833 |

16. $3.84 - 2.68 = n$

$\underline{\quad n = 1.16 \quad}$

17. $2.7 - 0.312 = n$

$\underline{\quad n = 2.388 \quad}$

18. $4.1 - 3.3 = n$

$\underline{\quad n = 0.8 \quad}$

19. $6.57 - 1.898 = n$

$\underline{\quad n = 4.672 \quad}$

20. $5.2 - 2.623 = n$

$\underline{\quad n = 2.577 \quad}$

21. $7.42 - 3.416 = n$

$\underline{\quad n = 4.004 \quad}$

Problem Solving Skill

Estimate or Find Exact Answer

Both estimation and exact answers are useful when shopping.

Estimations help you determine if you have enough money. Exact answers help you determine if you received the correct change.

Suppose you have $5.00, and want to buy 5 drinks for $0.85 each. Do you have enough money? How much change will you receive?

Estimation	$0.85	→	$1.00
	$0.85	→	$1.00
	$0.85	→	$1.00
	$0.85	→	$1.00
	$0.85	→	$1.00
			$5.00

Exact Answer	$0.85	$5.00
	$0.85	$-\ 4.25$
	$0.85	0.75 change
	$0.85	
	$0.85	
	$4.25	

So, you have enough money.   So, you should receive $0.75 change.

_____

Write an estimate of the total amount. Then solve. Estimates may vary.

1. Pat has $10.00. He wants to buy a magazine for $3.25, a small pizza for $3.89, and two drinks for $1.15 each. How much change will Pat receive?

   $9.00;

   $0.56

2. Paula has $45. She wants to purchase CDs costing $12.99, $14.99, $9.99, and $11.99. Does Paula have enough money? If so, how much change will she receive?

   $50.00; no

3. Jenny has $10.00. She wants to buy 6 pounds of apples costing $0.75 per pound and a bag of oranges costing $1.45. What is Jenny's exact cost? How much change will she receive?

   $7.00;

   $5.95; $4.05

4. Erin has $50.00. She wants to buy a purse for $17.99, gloves for $10.98, and a sweater for $19.95. What is her exact cost? How much change will Erin receive?

   $49.00;

   $48.92; $1.08

© Harcourt

Expressions and Variables

An expression has numbers and operation signs. It does not have an equal sign.

Use these words to help you write expressions.

Addition: more, sum, plus, added, gave

Subtraction: less, minus, loss, difference, spent, left

John had 12 marbles. He won 7 more.	Mary had $10. She spent $3.
Translate this into an expression.	Translate this into an expression.
Clue Word: <u>more</u> $12 + 7$	Clue Word: <u>spent</u> $10 - 3$

An expression may have a variable. A variable is a letter or symbol that can stand for a number.

Peter caught 2 fish in the morning. In the afternoon, he caught some more.	Susan had 4 sharpened pencils. Then she broke the point off of some of them.
Translate this into an expression.	Translate this into an expression.
Clue Word: more $2 + n$	Clue Word: left $4 - n$

Write the clues. Then write an expression using n for the unknown number. Explain what the variable represents.

1. The temperature dropped 7 degrees and then went up 4 degrees.
   Clue Words: dropped, went up;
   $n - 7 + 4$; $n =$ the beginning temperature in degrees

2. When the train stopped, 5 people boarded and 2 got off.
   Clue Words: boarded, got off;
   $n + 5 - 2$; $n =$ people on the train to start with

3. Steven wrote 8 pages for homework. The dog ate some of them.
   Clue Words: ate; $8 - n$;
   $n =$ the pages the dog ate

4. Gabriel collected 7 stones. John gave some more stones.
   Clue Words: gave; $7 + n$;
   $n =$ the stones John gave

Write Equations

An equation is a number sentence that uses the equal sign to show that two amounts are equal.

You can use variables to stand for numbers you do not know.

Peter had 10 books. After his birthday party, he had 16 books. How many books did he receive for his birthday?

books he has plus books received = total books

10 books + books received = total books

$$10 + n = 16$$

Write an equation with a variable. Explain what the variable represents.

1. Joseph had 7 paper cups. There were 22 students in the class. How many more cups did he need to serve punch to all his classmates?

cups he had + cups he needed = total cups for punch

$7 + n = 22;$

$n =$ the cups needed

2. Mary Beth loves chocolate chip cookies. Her mother took a sheet of 12 out of the oven. Mary Beth ate some. Now there are 8 left. How many did she eat?

total cookies − number eaten = number left

$12 - n = 8;$

$n =$ the number she ate

3. Jennifer had spent $32 for a new jacket. She had $12 left. How much did she have originally?

original amount − amount spent = amount left

$n - 32 = 12;$

$n =$ the original amount

4. Monica had a collection of stickers. She bought 7 and had a total of 21. How many did she originally have?

number in collection + number gained = total amount

$n + 7 = 21;$

$n =$ the number of the original collection

Solve Equations

When you solve an equation, you find the value of the variable that makes the equation true.

In an equation, the amounts on both sides of the equal sign have the same value. It is like a balanced scale.

$n + 6 = 10$

To solve, ask, "How many counters would I need to add to the left side of the scale to make it balanced?" Use mental math to find the missing addend.

The solution equation will be

$n + 6 = 10$.   Think: what number plus 6 equals 10?

$n = 4$

Check your solution. Replace n with 4.

$n + 6 = 10$

$4 + 6 = 10$

$10 = 10$

Use mental math to solve. Check your solution.

1. $n + 5 = 15$ **2.** $n - 6 = 6$ **3.** $n - 10 = 20$

_____ $n = 10$ _____   _____ $n = 12$ _____   _____ $n = 30$ _____

Solve the equation. Check your solution.

4. $15 + n = 22$ **5.** $n - 8 = 12$ **6.** $25 - n = 22$

_____ $n = 7$ _____   _____ $n = 20$ _____   _____ $n = 3$ _____

7. $n + 10 - 6 = 7$ **8.** $22 - n + 7 = 18$ **9.** $14 - 8 + n = 13$

_____ $n = 3$ _____   _____ $n = 11$ _____   _____ $n = 7$ _____

Use Addition Properties

You can use the properties of addition to help you solve problems.

The **Associative Property** states that you may group addends differently without changing the value of the sum.

$$7 + (8 + 4) = (7 + 8) + 4$$
$$7 + 12 = 15 + 4$$
$$19 = 19$$

The **Commutative Property** states that addends may be added in any order without changing the value of the sum.

$$6 + 5 = 5 + 6$$
$$11 = 11$$

The **Zero Property** states that you may add zero to any number without changing the value of the number.

$$5 + 0 = 5$$

Name the addition property used in each equation.

1. $223 + 0 = 223$ **2.** $(5 + 6) + 3 = 5 + (6 + 3)$ **3.** $56.4 + 10 = 10 + 56.4$

_____Zero Property_____   _Associative Property_   _Commutative Property_

Find the value of n. Identify the addition property used.

4. $200 + n = 100 + 200$ **5.** $78 + (5 + n) = (78 + 5) + 7$

n = 100; Commutative Property   _n = 7; Associative Property_

6. $4 + n = 7 + 4$ **7.** $0 + 88 = n$

n = 7; Commutative Property   _n = 88; Zero Property_

Algebra: Name the addition property used in each equation.

8. $g + h = h + g$ **9.** $p + (q + r) = (p + q) + r$

Commutative Property   _Associative Property_

10. $w + 0 = w$ **11.** $d + f = f + d$

_____Zero Property_____   _Commutative Property_

Problem Solving Skill

Use a Formula

To find the perimeter of a figure, you add the lengths of its sides. Remember that perimeter is the distance around a figure.

You can use a formula to find the perimeter. Use a different letter for each side of the figure.

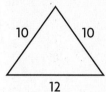

$P = a + b + c$

$P = 10 + 10 + 12$

$P = 32$

Find the perimeter of the following figures.

1.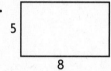

_____ 26 units _____

2.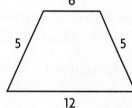

_____ 28 units _____

3.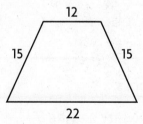

_____ 64 units _____

4.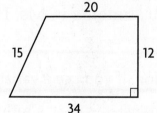

_____ 81 units _____

Use a formula to solve.

5. Jeff wants to build a rectangular fence in his yard for his dog. The yard is 35 feet by 40 feet. How much fencing must Jeff buy?

_____ $P = a + b + c + d,$ _____

_____ $P = 35 + 40 + 35 + 40,$ _____

_____ $P = 150$ feet _____

6. Draw a square with each side measuring 8 units and find the perimeter.

Check students' drawings.

_____ $8 + 8 + 8 + 8 = 32$ _____

Write and Evaluate Expressions

You can **write and evaluate expressions** to model different situations.

Ms. Hartwick has 6 rows of students in her classroom. She has the same number of students in each row.

To model how many students are in Ms. Hartwick's class, you can write an expression.

6 rows	times	number of students in each row
↓	↓	↓
6	×	n

If there are 5 students in each row, how many students are in Ms. Hartwick's class altogether?

Replace the variable n in the expression with 5 to find how many students are in Ms. Hartwick's class altogether.

$6 \times n$ Evaluate the expression if $n = 5$.
 ↓
6×5 Replace n with 5.
 ↓
30

So, there are 30 students in Ms. Hartwick's class altogether.

Write an expression. If you use a variable, tell what it represents.

1. Beth runs 4 days a week. She runs the same number of miles each day.

$4 \times m$; m = number of miles

2. Caitlin bought 12 boxes of canned dog food. Each box had 9 cans of dog food.

12×9

3. Marcus has 7 shelves of CDs. Each shelf holds the same number of CDs.

$7 \times c$; c = number of CDs

Evaluate each expression.

4. $n \times 2$ if $n = 12$

24

5. $9 \times n$ if $n = 7$

63

6. $22 + (n \times 3)$ if $n = 6$

40

Order of Operations

When an expression has more than one operation, you evaluate it using the order of operations. The order of operations is a set of rules that tells you which operation to do first.

Evaluate $18 + (4 \times 6) \div 2$.

Step 1	Operate inside <u>parentheses</u> .	$18 + (4 \times 6) \div 2$
		$\{ 4 \times 6 = 24 \}$
Step 2	<u>Multiply and divide</u> from left to right.	$18 + 24 \div 2$
		$\{ 24 \div 2 = 12 \}$
Step 3	<u>Add and Subtract</u> from left to right.	$18 + 12$
		$\{ 18 + 12 = 30 \}$

So, $18 + (4 \times 6) \div 2 = 30$

30

Complete to evaluate the expression.

1. $10 + (7 \times 4) - 8$

$10 + \underline{\quad 28 \quad} - 8$

$\underline{\quad 38 \quad} - 8$

$\underline{\quad 30 \quad}$

2. $15 \div 5 \times 9 - 4$

$\underline{\quad 3 \quad} \times 9 - 4$

$\underline{\quad 27 \quad} - 4$

$\underline{\quad 23 \quad}$

Evaluate the expression.

3. $14 - (5 + 2) \times 2$

$\underline{\qquad 0 \qquad}$

4. $2 \times 8 + (16 \div 4)$

$\underline{\qquad 20 \qquad}$

5. $5 \times 7 - 24 \div 8$

$\underline{\qquad 32 \qquad}$

6. $4 + (55 \div 11) \times 6$

$\underline{\qquad 34 \qquad}$

7. $29 - (6 \times 3) \div 2$

$\underline{\qquad 20 \qquad}$

8. $(27 \div 9) \times 8 + 7$

$\underline{\qquad 31 \qquad}$

9. $8 \times 6 - 7 \times 2$

$\underline{\qquad 34 \qquad}$

10. $30 - (10 \div 10) + 13$

$\underline{\qquad 42 \qquad}$

11. $6 \times 7 - 4 \times 5$

$\underline{\qquad 22 \qquad}$

12. $19 - 7 \times (12 \div 6)$

$\underline{\qquad 5 \qquad}$

13. $7 + 48 \div (7 + 5)$

$\underline{\qquad 11 \qquad}$

14. $27 \div 3 - (1 \times 5)$

$\underline{\qquad 4 \qquad}$

Functions

When one quantity depends on another quantity, the relationship between the quantities is called a function.

Paintbrushes cost $4 each. How much will 5 paintbrushes cost?

You can write an equation to represent the function.

number of dollars	=	4	×	number of paintbrushes
d	=	4	×	p
d	=	4	×	5
d	=	20		

You can also use a function table to show the number of dollars different numbers of paintbrushes cost.

paintbrushes, p	1	2	3	4	5
dollars, d	4	8	12	16	20

So, 5 paint brushes will cost $20.

Complete the function table.

1. $b = 9c$

c	2	4	6	8	10
b	18	36	54	72	90

2. $s = 7t$

t	6	7	8	9	10
s	42	49	56	63	70

3. $h = 6j + 4$

j	8	6	4	2	0
h	52	40	28	16	4

4. $f = 6 + 5g$

g	0	5	10	15	20
f	6	31	56	81	106

5. $d = 3a - 2$

a	12	10	8	6	4
d	34	28	22	16	10

6. $n = 15 + 2m - 4$

m	3	5	7	9	11
n	17	24	25	29	33

Use the function. Find the output, y for each input, x.

7. $y = 8x - 7$ for $x = 3, 4, 5$

17; 25; 33
_____

8. $y = 100 - 4x$ for $x = 5, 10, 20$

80; 60; 20
_____

9. $y = 6x + 15$ for $x = 6, 7, 8$

51; 57; 63
_____

10. $y = 49 - 3x$ for $x = 8, 9, 10$

25; 22; 19
_____

Problem Solving Strategy

Write an Equation

You can **write an equation** to help you solve a problem.

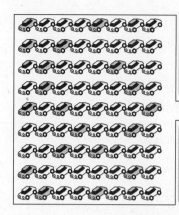

Felicity and Alex were in charge of parking cars in the small parking lot at the State Fair. The lot was filled with 72 cars in all by noon of the first day. The cars were organized into 9 equal rows of cars. How many cars were in each row?

Write an equation to find the number of cars parked in each row.

Think 9 times what number equals 72.

So, each row had 8 cars.

total cars = in lot		rows of cars	×	number of cars in each row
72	=	9	×	c
72	=	9	×	8
c	=	8		

Write and solve an equation for each problem. Explain what the variable represents. Possible answers are given.

1. Jacob has to stack boxes in the grocer's storage room. The room is 96 inches high. Each box is 12 inches high. How many boxes can Jacob stack on top of each other?

   $96 = 12 \times b$; b = number

   of boxes; 8

2. The shelves that the grocer stacks the canned goods on are 30 inches high. The grocer stacked the cans 5 high. How tall is each can?

   $30 = 5 \times h$; h = height of can;

   6 inches

3. Chelsea has to line up 48 chairs in 6 equal rows. How many chairs should she put in each row?

   $48 = 6 \times c$; c = number of

   chairs in each row; 8

4. Troy made a striped blanket for his bed. The blanket was 54 inches wide with 9 equal stripes. How wide was each stripe?

   $54 = 9 \times w$; w = width of each

   stripe; 6 inches

© Harcourt

Use Multiplication Properties

You can use mental math and the **properties of multiplication**
to solve problems.

Property of Multiplication	Example	Explanation
Commutative Property	$4 \times 2 = n \times 4$ $4 \times 2 = 2 \times 4$ $n = 2$	You can multiply numbers in any order. The product is always the same.
Associative Property	$(3 \times n) \times 5 = 3 \times (4 \times 5)$ $(3 \times 4) \times 5 = 3 \times (4 \times 5)$ $n = 4$	You can group factors differently. The product is always the same.
Property of One	$n \times 1 = 5$ $5 \times 1 = 5$ $n = 5$	When one of the factors is 1, the product equals the other number.
Zero Property	$4 \times n = 0$ $4 \times 0 = 0$ $n = 0$	When one factor is 0, the product is 0.

Solve the equation. Identify the property used.

1. $n \times 3 = 0$

$n = 0$; Zero

Property

2. $n \times 3 = 3 \times 2$

$n = 2$; Commutative

Property

3. $4 \times (2 \times 5) = (n \times 2) \times 5$

$n = 4$; Associative

Property

4. $1 \times n = 8$

$n = 8$; Property

of One

5. $(n \times 3) \times 2 = 5 \times (3 \times 2)$

$n = 5$; Associative

Property

6. $6 \times 7 = 7 \times n$

$n = 6$; Commutative

Property

7. $(7 \times 3) \times n = 7 \times (3 \times 2)$

$n = 2$; Associative

Property

8. $8 \times 2 = n \times 8$

$n = 2$; Commutative

Property

9. $3 \times n = 3$

$n = 1$; Property

of One

Name _____

The Distributive Property

You can use the **Distributive Property** to break apart numbers to make them easier to multiply.

To find 20 × 13, you can break apart 13.

20 × 13 = 20 × (10 + 3)   ← Break apart.

= (20 × 10) + (20 × 3) ← Multiply.

= (200) + (60)   ← Add.

= 260

Use the Distributive Property to restate each expression. Find the product. Check students' work.

1. 20 × 12

Break apart. 20 × (__10__ + __2__)

Multiply.   20 × __10__ = __200__

20 × __2__ = __40__

Add.   200 + __40__ = __240__

__240__

2. 20 × 18

Break apart. 20 × (__10__ + __8__)

Multiply.   20 × __10__ = __200__

20 × __8__ = __160__

Add.   __200__ + __160__ = __360__

__360__

3. 30 × 16

Break apart. 30 × (__10__ + __6__)

Multiply.   30 × __10__ = __300__

30 × __6__ = __180__

Add.   __300__ + __180__ = __480__

__480__

4. 12 × 45

Break apart. 12 × (__40__ + __5__)

Multiply.   12 × __40__ = __480__

12 × __5__ = __60__

Add.   __480__ + __60__ = __540__

__540__

5. 30 × 26

Break apart. 30 × (__20__ + __6__)

Multiply.   30 × __20__ = __600__

30 × __6__ = __180__

Add.   __600__ + __180__ = __780__

__780__

6. 25 × 17

Break apart. 25 × (__10__ + __7__)

Multiply.   25 × __10__ = __250__

25 × __7__ = __175__

Add.   __250__ + __175__ = __425__

__425__

© Harcourt

Collect and Organize Data

The tally table shows how many fifth grade students rode the bus during the first four weeks of school. How can you find the total number of students in the fifth grade that rode a bus to school?

Week	Fifth Grade Riders
1	~~HHH~~ //
2	~~HHH~~ /
3	~~HHH~~ ~~HHH~~
4	~~HHH~~

The information in the tally table can be made easier to read and understand by using a frequency table. The **frequency** for each week tells how many fifth grade students rode a bus that week. The **cumulative frequency** column shows a running total of the number of students who rode a bus. Check students' work.

Step 1 Count the tally marks for each week. Place the total for each week in the column labeled Frequency on the frequency table.

Step 2 For each new line of data, write the sum of the frequencies in the Cumulative Frequency column. The last number in the Cumulative Frequency column will tell you the total number of fifth graders that rode a bus.

FIFTH GRADERS RIDING A BUS		
Week	Frequency (Number of Students)	Cumulative Frequency
1	7	7
2	6	$7 + 6 = 13$
3	10	$13 + 10 = 23$
4	5	$23 + 5 = 28$

How many fifth graders rode a bus? ___28___

The **range** is the difference between the greatest and the least numbers in a set of data.   Greatest Number − Least Number = Range
Use the frequency table to find the range of the number of fifth graders who rode a bus. Show your work. _____ $10 - 5 = 5$ _____

Suppose 2 more fifth graders rode a bus in Week 3. In addition, 7 new fifth graders enrolled in school. 4 of the new students are walkers and 3 rode a bus in Week 2. Use this information to complete a new frequency table. What is the new total number of fifth graders riding a bus? What is the new range?

___ 29; 4 ___

FIFTH GRADERS RIDING A BUS		
Week	Frequency (Number of Students)	Cumulative Frequency
1	7	7
2	9	$7 + 9 = 16$
3	8	$16 + 8 = 24$
4	5	$24 + 5 = 29$

Find the Mean

Tom has taken three tests. He wants to know his average score for the three tests. The type of average Tom is looking for is called the **mean.**

Tom's Test Scores			
Test	1	2	3
Score	80	70	90

Step 1

Add the three test scores together.

$80 + 70 + 90 = 240$

Step 2

Divide the sum by the number of tests.

$240 ÷ 3 = 80$

So, Tom's mean test score is 80.

Write an addition sentence for the sum of each set of numbers.

1. 3, 5, 4, 1, 7   **2.** 20, 15, 10   **3.** 22, 26, 28, 32

$3 + 5 + 4 + 1 + 7 = 20$ $20 + 15 + 10 = 45$ $22 + 26 + 28 + 32 = 108$

Write how many numbers are listed in each set of numbers.

4. 3, 5, 4, 1, 7   **5.** 20, 15, 10   **6.** 22, 26, 28, 32

5   3   4

Write a division sentence to find the mean for each set of numbers.

7. 3, 5, 4, 1, 7   **8.** 20, 15, 10   **9.** 22, 26, 28, 32

$20 ÷ 5 = 4$ $45 ÷ 3 = 15$ $108 ÷ 4 = 27$

10. One month later, Tom took 5 more tests. His scores were 80, 70, 90, 90, and 100. What is the mean of these test scores? Show your work.

$80 + 70 + 90 + 90 + 100 = 430; 430 ÷ 5 = 86$

Find the mean for each set of data.

11. 9, 11, 13, 13, 9   **12.** 33, 28, 35, 33, 26   **13.** 105, 112, 133, 118, 102

11   31   114

Find the Median and Mode

Sam takes tests to see how many words he can type in a minute. The data in the table show his first 7 tests.

Number of Words Typed in a Minute							
Test	1	2	3	4	5	6	7
Score	22	16	18	14	16	34	20

You can find Sam's median score and the mode of the data.

Step 1

List the scores from least to greatest.

14, 16, 16, 18, 20, 22, 34

Step 2

To find the median score, cross off a number from each end until there is only one number left in the middle.

~~14~~, ~~16~~, ~~16~~, (18,) ~~20~~, ~~22~~, ~~34~~

The number 18 is the **median** score.

Step 3

Find the score that occurred most often.

Sam scored 16 twice.

The number 16 is the **mode**.

Sometimes there is more than one mode or no mode.

Arrange the numbers from least to greatest. Circle the median number.

1. 13, 12, 11, 11, 9, 8, 16, 17, 19

   8, 9, 11, 11, (12,) 13, 16, 17, 19

2. 24, 32, 28, 45, 19, 23, 16, 51, 32

   16, 19, 23, 24, (28,) 32, 32, 45, 51

3. 103, 98, 105, 101, 99

   98, 99, (101,) 103, 105

Arrange the numbers from least to greatest. Find the median and the mode.

4. 9, 7, 5, 11, 11

   5, 7, 9, 11, 11

   median: __9__

   mode: __11__

5. 14, 12, 12

   12, 12, 14

   median: __12__

   mode: __12__

6. 3, 7, 2, 9, 6, 5, 3, 1, 3

   1, 2, 3, 3, 3, 5, 6, 7, 9

   median: __3__

   mode: __3__

Problem Solving Strategy

Make a Graph

Mr. Schwartz recorded the number of newspapers he sold in his store every day of the week for two weeks. Newspapers sales were 60, 65, 66, 71, 71, 72, 74, 75, 76, 77, 79, 80, 81, and 83. Is the number sold usually in the 60's, 70's, or 80's?

You can make a stem-and-leaf plot to organize the data by place value.

Make a column of the tens digits of the data, listing them in order from least to greatest. These are the **stems.**

Stem	Leaves
6	
7	
8	

Beside each tens digit, record the ones digits of the data, in order from least to greatest. These are the **leaves.**

Stem	Leaves
6	0 5 6
7	1 1 2 4 5 6 7 9
8	0 1 3

The stem-and-leaf plot shows the greatest number of leaves are on the 7 stem. So, the number of newspapers sold is usually in the 70's.

Make a graph to solve. Check students' stem-and-leaf plots.

1. Lynnette's golf scores are 72, 74, 74, 78, 80, 82, 83, 87, 88, and 91. Does she usually score in the 70's, 80's, or 90's?

Stem	Leaves
7	2 4 4 8
8	0 2 3 7 8
9	1

   _____ 80's _____

2. The coach of the Tigers recorded the number of parents that attended each home baseball game. Parents' attendance was 16, 17, 23, 24, 29, 30, 33, 36, 36, and 38. Is parents' attendance usually in the 10's, 20's, or 30's?

Stem	Leaves
1	6 7
2	3 4 9
3	0 3 6 6 8

   _____ 30's _____

Analyze Graphs

Graphs help you to draw conclusions, answer questions, and make predictions about the data. Study the following graphs to answer the questions.

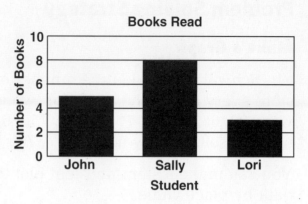

Books Read

1. A **bar graph** is useful when comparing data by groups.

   Which student read the most books? the least?

   _____Sally; Lori_____

2. **Line graphs** are helpful to see how data changes over a period of time.

   What happened to the temperature as the week passed?

   _____Possible answer:_____

   __In general, it got warmer__

   ____as the week passed.____

Daily Temperatures

3. A **circle graph** shows how parts of data relate to each other and to the whole.

   About one half of the animals in the pet store are what type of animal?

   _____birds_____

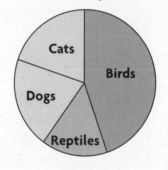

Pet Store Population

4. A **pictograph** displays countable data with symbols or pictures. Pictographs have a key to show how many each picture represents.

   How many books does Mr. Williams have in his class? ____56 books____

Types of Books in Mr. Williams' Class

Fantasy	▯▯▯▯▯ ▫
Mystery	▯▯▯
Biography	▯▯
Poetry	▯▯ ▫

Key: ▯ = 4 books

© Harcourt

Choose a Reasonable Scale

Henry kept track of how much mail his family received in one week.

He put the data in a table.

He wants to put the data in a line graph. He must select a scale. A **scale** is the series of numbers placed at fixed distances. The difference between one number and the next on the scale is called the **interval**.

Mail Received in a Week					
Day	Mon	Tue	Wed	Thu	Fri
Number of Pieces	8	10	6	4	2

The scale must include the numbers 2 through 10. It must include a number less than the least data and a number greater than the greatest data. Look at four ways Henry can display the mail data.

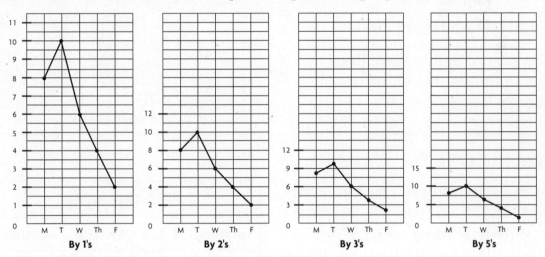

Henry selects a scale with intervals of 2.

From the box, choose the most reasonable interval for each set of data. List the numbers needed in the scale.

Interval
a. By 25's
b. By 20's
c. By 10's
d. By 5's

1. 5, 15, 20, 25, 10, 18

   d; 0, 5, 10, 15, 20, 25, 30

   _____

2. 50, 125, 100, 150, 100, 20

   a; 0, 25, 50, 75, 100,

   125, 150, 175

3. 8, 12, 10, 20, 10, 30

   c; 0, 10, 20, 30, 40

   _____

4. 20, 101, 40, 59, 115

   b; 0, 20, 40, 60, 80,

   100, 120

Problem Solving Strategy:

Make a Graph

The school population has changed over the last five years. Sid wants to use this data to predict next year's school population. He organized the data into a table.

Milton Elementary School Population					
Year	1997	1998	1999	2000	2001
Number of Students	450	520	560	580	620

Then he planned how to display the data using a bar graph. The interval skipped from 0 to 450, so Sid used a zigzag line to show a break in the scale.

He finds the range is 170.

He chooses the interval of 50.

Using the graph, Sid predicts that next year's population will be about 650 students.

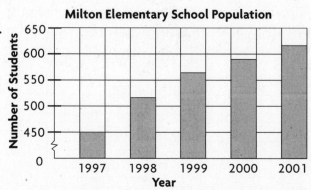

Make a graph to solve. Check students' graphs or plots.

1. Mr. Struther surveyed some students to find ideas for a field trip. He organized the data into a table. What graph or plot should he use to display the data? Make a graph or plot.

FIELD TRIP IDEAS			
Locations	Zoo	Museum	Aquarium
Number of Students	40	30	50

2. Attendance at the zoo was organized into a table. What graph or plot would best display the data? Make a graph or plot.

ATTENDANCE AT THE ZOO				
Month	April	May	June	July
Number of People	640	620	680	600

3. A baseball team kept track of the number of parents at the baseball games. The team organized the data into a table. What graph or plot would best display the data? Make a graph or plot.

PARENTS AT BASEBALL GAMES							
Game	1	2	3	4	5	6	7
Number of Parents	23	12	24	17	29	16	21

© Harcourt

Graph Ordered Pairs

Points on a coordinate grid can be given a unique name in the same way each house on a street has a unique number. Houses on a street follow an order so people can tell them apart and points also follow an order.

The order of the numbers in an ordered pair is always expressed the same way. The first number in an ordered pair tells how far to move horizontally from the origin. The second number tells how far to move vertically.

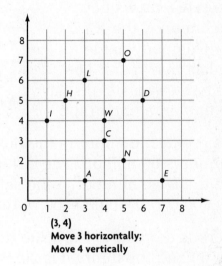

(3, 4)
Move 3 horizontally;
Move 4 vertically

Name the ordered pair for each point.

1. E _(7, 1)_   2. H _(2, 5)_

3. O _(5, 7)_   4. C _(4, 3)_

5. A _(3, 1)_   6. D _(6, 5)_

7. N _(5, 2)_   8. I _(1, 4)_

9. W _(4, 4)_   10. L _(3, 6)_

Graph and label the following points on a coordinate grid.
Check students' graphs.

11. M **(5, 7)**   12. N **(0, 5)**   13. P **(3, 4)**   14. R **(1, 0)**

15. S **(6, 2)**   16. A **(2, 5)**   17. V **(4, 1)**   18. G **(3, 7)**

19. B **(6, 0)**   20. H **(2, 6)**   21. T **(1, 7)**   22. Y **(6, 3)**

Make Line Graphs

The table shows how much money the XYZ Toy Company made for the last five years. The company wants to display the data in a line graph.

XYZ Toy Company					
Year	1995	1996	1997	1998	1999
Sales in Millions ($)	120	60	80	140	160

The greatest number in the table is 160 million. The least number is 60 million. The difference between the greatest and the least number in the set of data is the **range**.
160,000,000 − 60,000,000 = 100,000,000

There is a break in the scale from 0 to 60 and the interval is 20.

The vertical axis is labeled with the amounts of money; the horizontal axis is labeled with the years.

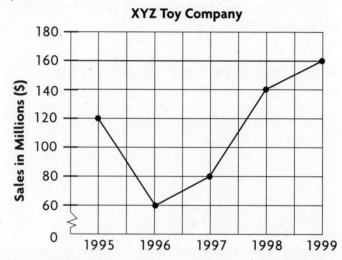

XYZ Toy Company

For each set of data, write a subtraction sentence to find the range.

1. 12, 18, 9, 13, 7

   18 − 7 = 11

2. 20, 100, 40, 60, 35

   100 − 20 = 80

3. 1, 9, 6, 5, 3

   9 − 1 = 8

Make a line graph for each set of data. Check students' graphs.

4.

On-Line Hours Used				
Month	Oct	Nov	Dec	Jan
Hours Used	30	60	100	125

5.

Jelly Bean Sales					
Days	1	2	3	4	5
Boxes Sold	100	50	150	250	200

Histograms

Histograms are a type of bar graph. The bars in a histogram are related and follow an order. They show the number of times the data occur within intervals.

The bars in a bar graph are not related to one another. To decide whether to make a histogram or bar graph, you need to decide whether the data fall within intervals.

Every 30 minutes, the popcorn popper records the number of boxes sold in the movie theater. Here is the information.

5:00	5:30	6:00	6:30	7:00	7:30	8:00	8:30	9:00	9:30	10:00	10:30	11:00
28	26	33	45	56	86	85	57	25	35	48	32	21

Find the range for the set of data. Range: 86 − 21 = 65; so the range is 65.

What interval could you use to make the histogram? Select an interval to divide the data equally. 13 intervals of 30 minutes

Make a frequency table with the intervals and record the number of boxes of popcorn sold during these time periods.

Use the frequency table to make the histogram. Label the scale for the number of boxes sold and title the graph. Graph the frequency for each interval.

Remember that the bars in a histogram are side-by-side.

Decide which graph would better represent the data below, a bar graph or histogram. Then make the graph. histogram; Check students' graphs.

MONEY SPENT ON LUNCH	
Amount of Money	**Number of Students**
$1.25	4
$1.75	3
$2.00	5
$2.50	4
$2.75	3
$3.00	1

Choose the Appropriate Graph

To display data, it is important to select the most appropriate graph or plot.

Two different graphs and two different plots for displaying data are shown.

Michael's Test Scores					
Test	1	2	3	4	5
Score	87	75	92	95	100

Line Plot

A line plot is used to record data as they are collected.

```
 x              x      x  x        x
─┼──────────────┼──────┼──┼────────┼─
 75             87     92 95       100
```

Stem-and-Leaf Plot

A stem-and-leaf plot is used to organize data by place value.

stem	leaf
7	5
8	7
9	2 5
10	0

Key: 7│5 = a score of 75

Bar Graph

A bar graph is used to compare facts about groups.

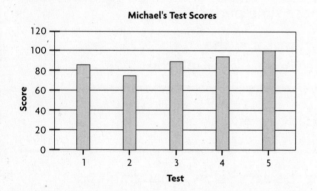

Line Graph

A line graph shows change over time.

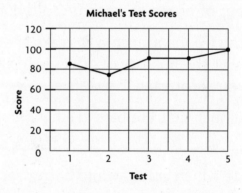

Michael chose the line graph because it shows his test scores over time.

Write the best type of graph or plot for the data.

Possible answers are given.

1. Compare the population of 6 major cities.

   _____ bar graph _____

2. Record the letter grades (A–F) for a class of 30 students.

   _____ line plot _____

3. Display the daily growth of a sunflower in inches.

   _____ line graph _____

4. Record the temperature every hour for 24 hours.

   _____ line plot _____

Estimation: Patterns in Multiples

You can round numbers and use basic facts to estimate products. Count the number of zeros in your rounded numbers. They will appear to the right of your basic fact in your estimate.

For 2-digit numbers:

If the ones digit is 0–4,
    round down.

If the ones digit is 5–9,
    round up.

For example: Round 40–44 to 40.
            Round 45–49 to 50.

$$\begin{array}{r} 45 \rightarrow\ \ 50 \\ \times 42 \rightarrow \underline{\times 40} \\ 2{,}0\underline{00} \end{array} \Big\rangle\ 2\ \text{zeros}$$

For 3-digit numbers:

If the tens digit is 0–4,
    round down.

If the tens digit is 5–9,
    round up.

For example: Round 700–749 to 700.
            Round 750–799 to 800.

$$\begin{array}{r} 749 \rightarrow\ \ 700 \\ \times\ 44 \rightarrow \times\ \ \ 40 \\ \underline{\hspace{2cm}} \\ 28{,}0\underline{00} \end{array} \Big\rangle\ 3\ \text{zeros}$$

Round each factor and estimate the product.

1. $\begin{array}{r} 141 \rightarrow\ 100 \\ \times\ 36 \rightarrow \times\ \ 40 \\ \hline 4{,}000 \end{array}$

2. $\begin{array}{r} 157 \rightarrow\ 200 \\ \times\ 57 \rightarrow \times\ \ 60 \\ \hline 12{,}000 \end{array}$

3. $\begin{array}{r} 125 \rightarrow\ 100 \\ \times\ 25 \rightarrow \times\ \ 30 \\ \hline 3{,}000 \end{array}$

4. $\begin{array}{r} 160 \rightarrow\ 200 \\ \times\ 41 \rightarrow \times\ \ 40 \\ \hline 8{,}000 \end{array}$

5. $\begin{array}{r} 187 \rightarrow\ 200 \\ \times\ 72 \rightarrow \times\ \ 70 \\ \hline 14{,}000 \end{array}$

6. $\begin{array}{r} 236 \rightarrow\ 200 \\ \times\ 45 \rightarrow \times\ \ 50 \\ \hline 10{,}000 \end{array}$

7. $\begin{array}{r} 349 \rightarrow\ 300 \\ \times\ 74 \rightarrow \times\ \ 70 \\ \hline 21{,}000 \end{array}$

8. $\begin{array}{r} 456 \rightarrow\ 500 \\ \times\ 56 \rightarrow \times\ \ 60 \\ \hline 30{,}000 \end{array}$

9. $\begin{array}{r} 568 \rightarrow\ 600 \\ \times\ 27 \rightarrow \times\ \ 30 \\ \hline 18{,}000 \end{array}$

10. $\begin{array}{r} 638 \rightarrow\ 600 \\ \times\ 16 \rightarrow \times\ \ 20 \\ \hline 12{,}000 \end{array}$

11. $\begin{array}{r} 774 \rightarrow\ 800 \\ \times\ 55 \rightarrow \times\ \ 60 \\ \hline 48{,}000 \end{array}$

12. $\begin{array}{r} 836 \rightarrow\ 800 \\ \times\ 43 \rightarrow \times\ \ 40 \\ \hline 32{,}000 \end{array}$

13. $\begin{array}{r} 719 \rightarrow\ 700 \\ \times\ 85 \rightarrow \times\ \ 90 \\ \hline 63{,}000 \end{array}$

14. $\begin{array}{r} 468 \rightarrow\ 500 \\ \times\ 68 \rightarrow \times\ \ 70 \\ \hline 35{,}000 \end{array}$

15. $\begin{array}{r} 229 \rightarrow\ 200 \\ \times\ 54 \rightarrow \times\ \ 50 \\ \hline 10{,}000 \end{array}$

Multiply by 1-Digit Numbers

Multiply the ones.   Multiply the tens.   Multiply the hundreds.

$$
\begin{array}{r}
14\boxed{3} \\
\times\ \ \boxed{3} \\
\hline
\boxed{9}
\end{array}
\qquad
\begin{array}{r}
1\boxed{4}3 \\
\times\ \ \boxed{3} \\
\hline
\boxed{2}9
\end{array}
\qquad
\begin{array}{r}
\boxed{1}43 \\
\times\ \ \boxed{3} \\
\hline
\boxed{4}29
\end{array}
$$

Sometimes you need to regroup.

Step 1 Multiply the ones. 3×3 ones = 9 ones

$$
\begin{array}{r}
143 \\
\times\ \ 3 \\
\hline
9
\end{array}
$$

Step 2 Multiply the tens. 3×4 tens = 12 tens
Write the 2. Regroup
the 10 tens as 1 hundred.

$$
\begin{array}{r}
^{1} \\
143 \\
\times\ \ 3 \\
\hline
29
\end{array}
$$

Step 3 Multiply the hundreds. 3×1 hundred = 3 hundreds
Now add the regrouped hundred.
3 hundreds + 1 hundred = 4 hundreds

So, $3 \times 143 = 429$.

$$
\begin{array}{r}
^{1} \\
143 \\
\times\ \ 3 \\
\hline
429
\end{array}
$$

Tell which place-value positions must be regrouped. Find the product.

1. $\begin{array}{r} 451 \\ \times\ \ 2 \\ \hline \end{array}$

 _____ tens: 902 _____

 _____

2. $\begin{array}{r} 328 \\ \times\ \ 3 \\ \hline \end{array}$

 _____ ones: 984 _____

 _____

3. $\begin{array}{r} 715 \\ \times\ \ 5 \\ \hline \end{array}$

 _____ ones, hundreds: _____

 _____ 3,575 _____

4. $\begin{array}{r} 1,458 \\ \times\ \ 6 \\ \hline \end{array}$

 _____ ones, tens, _____

 _____ hundreds: 8,748 _____

5. $\begin{array}{r} 2,473 \\ \times\ \ 2 \\ \hline \end{array}$

 _____ tens: 4,946 _____

 _____

6. $\begin{array}{r} 6,925 \\ \times\ \ 4 \\ \hline \end{array}$

 _____ ones, tens, hundreds, _____

 _____ thousands: 27,700 _____

7. $\begin{array}{r} 3,562 \\ \times\ \ 7 \\ \hline \end{array}$

 _____ ones, tens, hundreds, _____

 _____ thousands: 24,934 _____

8. $\begin{array}{r} 20,317 \\ \times\ \ 4 \\ \hline \end{array}$

 _____ ones, hundreds: _____

 _____ 81,268 _____

9. $\begin{array}{r} 13,234 \\ \times\ \ 3 \\ \hline \end{array}$

 _____ ones, tens: _____

 _____ 39,702 _____

Multiply by 2-Digit Numbers

You can multiply by two-digit numbers by breaking apart one of the factors.

To find 21 × 14, you can break apart 14 into 1 ten 4 ones.

Step 1 Multiply by the ones.

$$\begin{array}{r} 21 \\ \times 4 \\ \hline 84 \end{array}$$

Step 2 Multiply by the tens.

$$\begin{array}{r} 21 \\ \times 10 \\ \hline 210 \end{array}$$

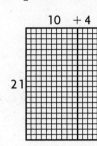

10 + 4

21

Step 3 Add the partial products.

$$\begin{array}{r} 21 \\ \times\,14 \\ \hline 84 \leftarrow 4 \times 21 \\ +210 \leftarrow 10 \times 21 \\ \hline 294 \end{array}$$

So, 21 × 14 = 294.

Complete to find the product.

1.
$$\begin{array}{r} 13 \\ \times 12 \\ \hline 26 \\ +130 \\ \hline 156 \end{array}$$
← __2__ × __13__
← __10__ × __13__

2.
$$\begin{array}{r} 22 \\ \times 15 \\ \hline 110 \\ +220 \\ \hline 330 \end{array}$$
← __5__ × __22__
← __10__ × __22__

3.
$$\begin{array}{r} 30 \\ \times 17 \\ \hline 210 \\ +300 \\ \hline 510 \end{array}$$
← __7__ × __30__
← __10__ × __30__

4.
$$\begin{array}{r} 28 \\ \times 14 \\ \hline 112 \\ +280 \\ \hline 392 \end{array}$$
← __4__ × __28__
← __10__ × __28__

5.
$$\begin{array}{r} 40 \\ \times 19 \\ \hline 360 \\ +400 \\ \hline 760 \end{array}$$
← __9__ × __40__
← __10__ × __40__

6.
$$\begin{array}{r} 45 \\ \times 15 \\ \hline 225 \\ +450 \\ \hline 675 \end{array}$$
← __5__ × __45__
← __10__ × __45__

7.
$$\begin{array}{r} 37 \\ \times 15 \\ \hline 185 \\ +370 \\ \hline 555 \end{array}$$
← __5__ × __37__
← __10__ × __37__

8.
$$\begin{array}{r} 28 \\ \times 16 \\ \hline 168 \\ +280 \\ \hline 448 \end{array}$$
← __6__ × __28__
← __10__ × __28__

Choose a Method

You can multiply three-digit numbers by breaking apart one of the factors.

To find 312 × 143, break apart 143 into 1 hundred 4 tens 3 ones.

Step 1	**Step 2**	**Step 3**	**Step 4**
Multiply by the ones.	Multiply by the tens.	Multiply by the hundreds.	Add the partial products.

<table>
<tr><td>

$$\begin{array}{r} 312 \\ \times \quad 3 \\ \hline 936 \end{array}$$

</td><td>

$$\begin{array}{r} 312 \\ \times \quad 40 \\ \hline 12{,}480 \end{array}$$

</td><td>

$$\begin{array}{r} 312 \\ \times \quad 100 \\ \hline 31{,}200 \end{array}$$

</td><td>

$$\begin{array}{r} 312 \\ \times \quad 143 \\ \hline 936 \\ 12{,}480 \\ 31{,}200 \\ \hline 44{,}616 \end{array}$$

936 ← 3 × 312
12,480 ← 40 × 312
31,200 ← 100 × 312

</td></tr>
</table>

So, 312 × 143 = 44,616.

Complete to find the product.

1.

$$\begin{array}{r} 423 \\ \times \quad 146 \\ \hline 2{,}538 \\ 16{,}920 \\ + 42{,}300 \\ \hline 61{,}758 \end{array}$$

2,538 ← __6__ × __423__
16,920 ← __40__ × __423__
42,300 ← __100__ × __423__

2.

$$\begin{array}{r} 231 \\ \times \quad 123 \\ \hline 693 \\ 4{,}620 \\ +23{,}100 \\ \hline 28{,}413 \end{array}$$

693 ← __3__ × __231__
4,620 ← __20__ × __231__
23,100 ← __100__ × __231__

3.

$$\begin{array}{r} 354 \\ \times \quad 246 \\ \hline 2{,}124 \\ 14{,}160 \\ +70{,}800 \\ \hline 87{,}084 \end{array}$$

← __6__ × __354__
← __40__ × __354__
← __200__ × __354__

4.

$$\begin{array}{r} 438 \\ \times \quad 253 \\ \hline 1{,}314 \\ 21{,}900 \\ +87{,}600 \\ \hline 110{,}814 \end{array}$$

1,314 ← __3__ × __438__
21,900 ← __50__ × __438__
87,600 ← __200__ × __438__

5.

$$\begin{array}{r} 672 \\ \times \quad 334 \\ \hline 2{,}688 \\ 20{,}160 \\ +201{,}600 \\ \hline 224{,}448 \end{array}$$

← __4__ × __672__
← __30__ × __672__
← __300__ × __672__

6.

$$\begin{array}{r} 596 \\ \times \quad 254 \\ \hline 2{,}384 \\ 29{,}800 \\ +119{,}200 \\ \hline 151{,}384 \end{array}$$

← __4__ × __596__
← __50__ × __596__
← __200__ × __596__

Problem Solving Skill

Evaluate Answers for Reasonableness

You can use estimation to check if an answer is reasonable.
Use your knowledge of patterns in multiples to help you with
large numbers.

At the garden center, there were 174 rows of flowers. Each row
contained 86 flowers. Estimate first. Then solve the problem and
compare it to your estimate to see if it is reasonable.

Estimate

$$
\begin{array}{r}
200 \\
\times\ 90 \\
\hline
18,000
\end{array}
$$

$$
\begin{array}{r}
174 \\
\times\ 86 \\
\hline
1,044 \\
+13,920 \\
\hline
14,964
\end{array}
$$

Your estimate was more than your exact amount because you used greater
numbers.

How would your estimate compare to your exact answer if you rounded both
factors down?
Possible answer: your estimate would be less than the exact answer.

Choose the most reasonable answer without solving.

1. Joel's dad sold each of his paintings at an art show for $750. He
   sold 26 at the show. About how
   much money did he get?

   A $7,000   C $20,000
   B $12,000   D $60,000

2. Joel's dad pays $157 for the
   materials to create each of his
   paintings. About how much
   does it cost him to create the
   26 paintings he sold?

   F $600   H $5,000
   G $1,500   J $60,000

3. A small airport has 21,795
   passengers each year. About how
   many passengers will they have
   altogether in 8 years?

   A 2,000   C 200,000
   B 20,000   D 2,000,000

4. The average length of important
   rivers in the world is 2,142 miles.
   If we measured 18 of these rivers,
   about how many miles would we
   measure?

   F 4,000 miles   H 40,000 miles
   G 20,000 miles   J 400,000 miles

Multiply Decimals and Whole Numbers

To multiply a whole number and a decimal, modeling with
money can be helpful. To multiply 3×0.2, follow these steps.

Step 1
Write 0.2 as 0.20. You can add a
zero at the end of a decimal without
changing the value. Draw that
amount of money.

The 2 dimes equal $0.20. You could
also draw 4 nickels or 20 pennies.

Step 2
Draw 3 groups of coins of $0.20.

Count the total amount.
$0.20 + $0.20 + $0.20 = $0.60
So, $3 \times 0.2 = 0.60$, or 0.6.

Draw the coins that equal the decimal amount. Use the fewest
coins possible.

1. 0.28

2. 0.30

3. 0.16

4. 0.52

5. 0.80

6. 0.24

Make a money model to find each product. Check students' models.

7. $4 \times 0.15 =$ ___0.60___   **8.** $3 \times 0.1 =$ ___0.3, or 0.30___   **9.** $2 \times 0.21 =$ ___0.42___

10. $4 \times 0.01 =$ ___0.04___   **11.** $3 \times 0.06 =$ ___0.18___   **12.** $2 \times 0.78 =$ ___1.56___

13. $3 \times 0.32 =$ ___0.96___   **14.** $4 \times 0.12 =$ ___0.48___   **15.** $2 \times 0.53 =$ ___1.06___

Algebra: Patterns in Decimal Factors and Products

You can use patterns to place the decimal point in a product.

Factors		Product
2 ×	1 =	2
2 ×	0.1 =	0.2
2 ×	0.01 =	0.02

2 × 1 = 2   ← no decimal places in factors

2 × 0.1 = 0.2   ← one decimal place in factors

2 × 0.01 = 0.02   ← two decimal places in factors

The number of decimal places in the factors equals the number of decimal places in the product.

Complete the tables.

1.

2 ×	3 =	6
2 ×	0.3 =	0.6
2 ×	0.03 =	0.06

2.

2 ×	4 =	8
2 ×	0.4 =	0.8
2 ×	0.04 =	0.08

3.

3 ×	3 =	9
3 ×	0.3 =	0.9
3 ×	0.03 =	0.09

4.

3 ×	5 =	15
3 ×	0.5 =	1.5
3 ×	0.05 =	0.15

5.

3 ×	6 =	18
3 ×	0.6 =	1.8
3 ×	0.06 =	0.18

6.

3 ×	7 =	21
3 ×	0.7 =	2.1
3 ×	0.07 =	0.21

7.

2 ×	8 =	16
2 ×	0.8 =	1.6
2 ×	0.08 =	0.16

8.

4 ×	5 =	20
4 ×	0.5 =	2.0
4 ×	0.05 =	0.20

9.

6 ×	7 =	42
6 ×	0.7 =	4.2
6 ×	0.07 =	0.42

10.

15 ×	1 =	15
15 ×	0.1 =	1.5
15 ×	0.01 =	0.15

11.

28 ×	1 =	28
28 ×	0.1 =	2.8
28 ×	0.01 =	0.28

12.

32 ×	1 =	32
32 ×	0.1 =	3.2
32 ×	0.01 =	0.32

Model Decimal Multiplication

To multiply 0.3×0.2, a 10-by-10 model will help.

Step 1: Draw diagonal lines through the bottom 3 rows.

The 3 rows represent 0.3.

Step 2: Draw diagonal lines through 2 columns.

The 2 columns represent 0.2.

Step 3: The overlapping squares that have an x in them show the product of 0.3×0.2.

The 6 squares with x's represent 0.06.

The product of 0.3×0.2 is 0.06.

Write a number sentence for each drawing.

1.
2.
3.
4.

$0.2 \times 0.4 = 0.08$ $0.4 \times 0.6 = 0.24$ $0.7 \times 0.2 = 0.14$ $0.6 \times 0.3 = 0.18$

Make a model for each and find the product.

5. $0.1 \times 0.5 = \underline{\quad 0.05 \quad}$

6. $0.2 \times 0.8 = \underline{\quad 0.16 \quad}$

7. $0.5 \times 0.9 = \underline{\quad 0.45 \quad}$

8. $0.7 \times 0.5 = \underline{\quad 0.35 \quad}$

Place the Decimal Point

How many decimal places are in the product of 0.21 and 0.03?

Step 1

Find the total number of decimal places in the factors.

$$0.21 \quad \times \quad 0.03 = \qquad ?$$

2 places + 2 places = 4 places

$$\begin{array}{r} 0.21 \\ \times\ 0.03 \\ \hline 0.\_\_\_\_ \end{array}$$

Step 2

Multiply the numbers just like whole numbers. To have 4 decimal places, you have to add 2 zeros before the 63.

$$\begin{array}{r} 0.21 \\ \times\ 0.03 \\ \hline 63 \\ +\ 000 \\ \hline 0.0063 \end{array}$$

Write how many decimal places are in each number.

1. 0.105

_____three_____

2. 0.0006

_____four_____

3. 0.008

_____three_____

Write how many decimal places are in each product. Then write the product. The first one has been done for you.

4. 0.3 × 0.5

____ = **0.** __ __

____ = **0.15**

5. 0.6 × 0.03

____ = 0. ____

____ = 0.018

6. 0.002 × 0.8

____ = 0. ____

____ = 0.0016

7. 0.24 × 0.01

____ = 0. ____

____ = 0.0024

8. 3 × 0.4

____ = 0. __

____ = 1.2

9. 0.7 × 0.2

____ = 0. __

____ = 0.14

Find each product.

10. 0.5 × 0.03 = __0.015__   **11.** 0.06 × 1.8 = __0.108__   **12.** 7 × 0.08 = __0.56__

Zeros in the Product

Be careful when multiplying by decimals to include all of the decimal places in the product.

Example: Find 0.013 × 0.6.

Step 1

Find the number of decimal places the product should have.

0.013 has three decimal places and 0.6 has one decimal place. The product should have 3 + 1 = 4 decimal places.

Step 2

Multiply.

$$\begin{array}{r} 0.013 \\ \times\ \ 0.6 \\ \hline 78 \end{array}$$

Step 3

Place the decimal point.

The product should have 4 decimal places. There are two digits, so write 2 zeros in the product and place the decimal point.

$$\begin{array}{r} 0.013 \\ \times\ \ 0.6 \\ \hline 0.0078 \end{array}$$

---

1. Find 0.03 × 0.4.

Step 1:
Find the number of decimal places the product should have.

3

Step 2:
Multiply

$$\begin{array}{r} 0.03 \\ \times\ 0.4 \\ \hline 12 \end{array}$$

Step 3:
Place the decimal point.

$$\begin{array}{r} 0.03 \\ \times\ 0.4 \\ \hline 0.012 \end{array}$$

2. Find 0.047 × 0.07.

Step 1:
5

Step 2:
$$\begin{array}{r} 0.047 \\ \times\ 0.07 \\ \hline 329 \end{array}$$

Step 3:
$$\begin{array}{r} 0.047 \\ \times\ 0.07 \\ \hline 0.00329 \end{array}$$

3. Find 0.0732 × 0.8.

Step 1:
5

Step 2:
$$\begin{array}{r} 0.0732 \\ \times\ 0.8 \\ \hline 5856 \end{array}$$

Step 3:
$$\begin{array}{r} 0.0732 \\ \times\ 0.8 \\ \hline 0.05856 \end{array}$$

4. Find 0.054 × 0.007.

0.000378

5. Find 0.0942 × 0.7.

0.06594

Problem Solving Skill

Make Decisions

We make decisions every day. There are often many things to consider. Use the questions below to guide you through making decisions.

Your neighbors have invited you to go with them on Saturday. Julia's family is going to the museum and to a movie. Karl's family is going on a bakery tour and to a football game. You must decide which invitation to accept.

1. If the museum visit will cost $3.00 and the movie will cost $4.75, how much will the trip with Julia's family cost?

   _____ $7.75 _____

2. If a football ticket costs $14.50 and the bakery tour is free, how much will the trip with Karl's family cost?

   _____ $14.50 _____

3. If you had to make your decision based on total cost, which trip would you choose? Why?

   _____ Possible answer: the trip with Julia's family; it costs less _____

   _____

4. Julia's family will start their trip at 8:30 A.M. Breakfast will take 45 minutes. They plan to stay at the museum for 2 hours. Lunch will take 45 minutes, and the movie will last 2 hours and 30 minutes. When will the trip with Julia's family end?

   _____ 2:30 P.M. _____

5. The bakery tour will take 1 hour and 30 minutes. Lunch will take 30 minutes. The football game will take 3 hours and 30 minutes, and dinner with Karl's family will take 1 hour. If this trip starts at 11:00 A.M., when will it end?

   _____ 5:30 P.M. _____

6. If you had to make your decision based on the total time of the trip, the start time, or the end time of the trip, which invitation would you accept? Why?

   _____ Check students' answers. _____

   _____

7. If you had to make your decision based on the activities you like better, which invitation would you accept? Why?

   _____ Check students' answers. _____

   _____

© Harcourt

Estimate Quotients

Compatible numbers are numbers that are easy to compute
mentally. One compatible number divides evenly into the other.
Think of basic facts to find compatible numbers.

What is $8\overline{)554}$?

Step 1

Think: What are the multiples of 8?

8 16 24 32
40 48 **56** 64

Which multiple is closest to 55?
56 is close to 55.
8 and 560 are compatible numbers.

Step 2

Divide.

$560 \div 8 = 70$
A good estimate for $554 \div 8$ is 70.

Follow the steps above to estimate each quotient. Estimates may vary.

1. $3\overline{)252}$

$240 \div 3 = 80$

_____

2. $6\overline{)546}$

$540 \div 6 = 90$

_____

3. $4\overline{)154}$

$160 \div 4 = 40$

_____

4. $9\overline{)192}$

$180 \div 9 = 20$

_____

5. $7\overline{)129}$

$140 \div 7 = 20$

_____

6. $4\overline{)265}$

$280 \div 4 = 70$

_____

7. $8\overline{)344}$

$320 \div 8 = 40$

_____

8. $5\overline{)480}$

$500 \div 5 = 100$

_____

9. $2\overline{)497}$

$500 \div 2 = 250$

_____

10. $3\overline{)287}$

$270 \div 3 = 90$

_____

11. $5\overline{)6,558}$

$6,500 \div 5 = 1,300$

_____

12. $6\overline{)5,097}$

$5,400 \div 6 = 900$

_____

© Harcourt

Divide 3-Digit Dividends

Bryan has 522 coins. He divides them among 3 jars. How many coins are in each jar?

Divide. $522 \div 3 = n$

Step 1

Since 5 hundreds can be divided by 3, the first digit is in the hundreds place.
Divide. $3\overline{)5}$
    Multiply. 3×1
    Subtract. $5 - 3$
    Compare. $2 < 3$

$$\begin{array}{r} 1 \\ 3\overline{)522} \\ -\underline{3} \\ 2 \end{array}$$

Step 2

Bring down the tens. Divide. $3\overline{)22}$
    Multiply. 3×7
    Subtract. $22 - 21$
    Compare. $1 < 3$

$$\begin{array}{r} 17 \\ 3\overline{)522} \\ -\underline{3} \\ 22 \\ -\underline{21} \\ 1 \end{array}$$

Step 3

Bring down the ones. Divide. $3\overline{)12}$
    Multiply. 3×4
    Subtract. $12 - 12$
    Compare. $0 < 3$

$$\begin{array}{r} 174 \\ 3\overline{)522} \\ -\underline{3} \\ 22 \\ -\underline{21} \\ 12 \\ -\underline{12} \\ 0 \end{array}$$

Since $n = 174$, each jar contains 174 coins.

Follow the steps above to find each quotient.

1. $3\overline{)928}$ 309 r1

2. $7\overline{)149}$ 21 r2

3. $5\overline{)845}$ 169

4. $4\overline{)892}$ 223

5. $6\overline{)399}$ 66 r3

6. $3\overline{)873}$ 291

7. $9\overline{)765}$ 85

8. $5\overline{)934}$ 186 r4

Zeros in Division

There are 618 pencils in the supply room. They are to be divided evenly among 6 classes. How many pencils will each class receive?

You will use division to find the answer. $618 \div 6 = n$

Step 1

Since 6 hundreds can be divided by 6, the first digit will be in the hundreds place. Divide.

$$\begin{array}{r} 1 \\ 6\overline{)618} \\ -6 \\ \hline 0 \end{array}$$

Multiply.
$6 \times 1 = 6$
Subtract.
$6 - 6 = 0$
Compare.
$0 < 6$

Step 2

Bring down the tens. Divide the 1 ten. Since 6 >1, write 0 in the quotient.

$$\begin{array}{r} 10 \\ 6\overline{)618} \\ -6 \\ \hline 01 \\ -0 \\ \hline 1 \end{array}$$

Multiply.
$6 \times 0 = 0$
Subtract.
$1 - 0 = 1$
Compare.
$1 < 6$

Step 3

Bring down the ones. Divide.

$$\begin{array}{r} 103 \\ 6\overline{)618} \\ -6 \\ \hline 01 \\ -0 \\ \hline 18 \\ -18 \\ \hline 0 \end{array}$$

Multiply.
$6 \times 3 = 18$
Subtract.
$18 - 18 = 0$
Compare.
$0 < 6$

So, each class will receive 103 pencils.

Follow the steps above to find each quotient.

1. $3\overline{)927}$ 2. $8\overline{)872}$ 3. $5\overline{)542}$ 4. $6\overline{)608}$

5. $3\overline{)624}$ 6. $2\overline{)807}$ 7. $4\overline{)826}$ 8. $7\overline{)843}$

Choose a Method

Divide 42,574 by 7.

Divide 7 into 42 to get 6. Multiply 6 by 7 to get 42. Subtract 42 from 42 to get 0. Bring down the 5 to get 05.

Divide 7 into 5 to get 0; Multiply 0 by 7 to get 0. Subtract 0 from 05 to get 5. Bring down the 7 to get 57.

Divide 7 into 57 to get 8. Multiply 8 by 7 to get 56. Subtract 56 from 57 to get 1. Bring down the 4 to get 14.

Divide 7 into 14 to get 2. Multiply 2 by 7 to get 14. Subtract 14 from 14 to get 0.

So, 42,574 ÷ 7 = 6,082.

$$
\begin{array}{r}
6{,}082 \\
7\overline{)42{,}574} \\
-\,42\downarrow\quad \\
\overline{05}\quad\;\; \\
-\,0\downarrow\;\; \\
\overline{57}\;\; \\
-\,56\downarrow \\
\overline{14} \\
-\,14 \\
\overline{0}
\end{array}
$$

Follow the steps above to find each quotient.

1. 9)45,035   *5,003 r 8*

2. 5)9,085   *1,817*

3. 4)16,087   *4,021 r 3*

4. 5)70,861   *14,172 r 1*

5. 6)856,412   *142,735 r2*

6. 5)18,005   *3,601*

7. 4)200,088   *50,022*

8. 5)7,555   *1,511*

9. 5)654,321   *130,864 r1*

10. 5)21,076   *4,215 r 1*

11. 3)356,789   *118,929 r 2*

12. 3)67,530   *22,510*

13. 6)3,791   *631 r 5*

14. 7)4,326   *618*

15. 8)1,999   *249 r 7*

16. 3)645,123   *215,041*

Algebra: Expressions and Equations

An **expression** combines numbers or variables with operations.

Problem: twenty-four divided by a number   **Expression:** $24 \div n$

The **value** of the expression depends on n. If n is 2, the value is **12.** If n is 8, the value is **3.**

An **equation** is a number sentence that uses an equal sign to show that two amounts are equal.

Problem: Twenty-four divided by a number is six.   **Equation:** $24 \div n = 6$

To solve the equation, think: 24 divided by what number equals 6?

You can predict and test to solve.

Predict: 3   Test: $24 \div 3 = 8$; too high   Predict: 4   Test: $24 \div 4 = 6$; correct

---

Evaluate the expression for n.

1. $48 \div n$
   $n = 2, 6, 12$
   $n = 24, 8, 4$

2. $n \div 10$
   $n = 100, 60, 70$
   $n = 10, 6, 7$

3. $n \div 12$
   $n = 12, 36, 72$
   $n = 1, 3, 6$

4. $18 \div n$
   $n = 2, 6, 18$
   $n = 9, 3, 1$

Determine which value is a solution for the given equation.

5. $49 \div n = 7$
   $n = 5, 6, \text{ or } 7$
   $n = 7$

6. $195 \div n = 65$
   $n = 2, 3, \text{ or } 4$
   $n = 3$

7. $n \div 6 = 8$
   $n = 36, 42, \text{ or } 48$
   $n = 48$

8. $n \div 5 = 50$
   $n = 200, 250, \text{ or } 300$
   $n = 250$

9. $350 \div n = 50$
   $n = 5, 6, \text{ or } 7$
   $n = 7$

10. $200 \div n = 5$
    $n = 30, 40, \text{ or } 50$
    $n = 40$

11. $n \div 6 = 12$
    $n = 72, 76, \text{ or } 78$
    $n = 72$

12. $n \div 6 = 14$
    $n = 80, 82, \text{ or } 84$
    $n = 84$

Solve each equation. Then, check the solution.

13. $45 \div n = 5$
    $n = 9$

14. $100 \div n = 10$
    $n = 10$

15. $n \div 6 = 7$
    $n = 42$

16. $n \div 12 = 9$
    $n = 108$

17. $65 \div n = 13$
    $n = 5$

18. $120 \div n = 30$
    $n = 4$

19. $n \div 3 = 31$
    $n = 93$

20. $n \div 4 = 21$
    $n = 84$

© Harcourt

Name _____

Problem Solving Skill

Interpret the Remainder

When there is a remainder in a division problem, you need to look at the question to see what is being asked. You may drop the remainder, or round the quotient to the next greater whole number, or you may use the remainder as a fractional part of your answer.

Andy made punch with 48 ounces of apple juice, 36 ounces of grape juice, and 60 ounces of lemon soda. How many 5-ounce servings did he make?

$48 + 36 + 60 = 144$ oz

$$\begin{array}{r} 28\ r4 \\ 5\overline{)144} \\ -10 \\ \hline 44 \\ -40 \\ \hline 4 \end{array}$$

There are 4 ounces left over. That is not enough for another 5-ounce serving. Drop the remainder.

So, Andy made 28 five-ounce servings.

Solve. Explain how you interpreted the remainder.

1. Mia bought 10 feet of wire for a science project. She divided the wire equally into 3 pieces. How long was each piece of wire?

   $3\frac{1}{3}$ ft; Use the remainder as

   part of the answer.

2. A total of 325 people will be attending a sports banquet. There will be 8 people seated at each table. How many tables will be needed?

   41 tables; Round to the next

   greater whole number.

3. A total of 175 players signed up for a baseball league. There are 9 teams in the league. If the players are divided among the teams, what is the greatest number of players on any team?

   20 players; Round to the next

   greater whole number.

4. Jennie baked 132 cookies. She wants to divide them evenly among her 7 friends. How many cookies will she give to each friend?

   18 cookies; Drop the remainder.

Algebra: Patterns in Division

Rick has a 1,313-page book. If he reads 14 pages a day,
about how long will it take him to finish reading the book?
You divide to find the answer.

$1,313 \div 14$

You can estimate to find the number of days it will take Rick
to read the book.

Estimate: 1,313 rounds down to 1,000.
14 rounds down to 10.

$1,000 \div 10$

There are zeros in the dividend and in the divisor. Cancel
out one zero in each.

$1,00\cancel{0} \div 1\cancel{0} = 100$

So, it will take Rick about 100 days to read the book.

You can check this estimate by multiplying. Multiply the
divisor by the quotient.

$10 \times 100 = 1,000$

Find each quotient. Cancel out the zeros if appropriate. Write a
multiplication sentence to check. The first one is done for you.

1. $1,50\cancel{0} \div 3\cancel{0} = 50$

   $30 \times 50 = 1,500$

2. $560 \div 70 = \underline{\quad 8 \quad}$

   $70 \times 8 = 560$

3. $720 \div 80 = \underline{\quad 9 \quad}$

   $80 \times 9 = 720$

4. $2,100 \div 70 = \underline{\quad 30 \quad}$

   $70 \times 30 = 2,100$

5. $480 \div 60 = \underline{\quad 8 \quad}$

   $60 \times 8 = 480$

6. $2,500 \div 50 = \underline{\quad 50 \quad}$

   $50 \times 50 = 2,500$

7. $36,000 \div 90 = \underline{\quad 400 \quad}$

   $90 \times 400 = 36,000$

8. $24,000 \div 40 = \underline{\quad 600 \quad}$

   $40 \times 600 = 24,000$

9. $5,600 \div 80 = \underline{\quad 70 \quad}$

   $80 \times 70 = 5,600$

Estimate Quotients

Compatible numbers are numbers that are close to the actual numbers and can be divided evenly. They can help you estimate a quotient.

Estimate. $42\overline{)1{,}574}$

Step 1
Round the divisor.

The number 42 rounds to 40. It can also be rounded to 50.

Step 2
Round the dividend.

The number 1,574 can be rounded up to 1,600 or rounded down to 1,500.

Step 3
Rewrite the division problem with the compatible numbers, and solve.

$$40\overline{)1{,}600}^{\,40} \qquad\qquad 50\overline{)1{,}500}^{\,30}$$

So, one estimate of the quotient is 40. A second estimate is 30.

Write two pairs of compatible numbers for each. Give two possible estimates. Two possible pairs and their estimates are given.

1. $48\overline{)3{,}367}$

$3{,}500 \div 50 = 70$

$3{,}200 \div 40 = 80$

2. $76\overline{)4{,}117}$

$4{,}000 \div 80 = 50$

$4{,}200 \div 70 = 60$

3. $37\overline{)847}$

$800 \div 40 = 20$

$900 \div 30 = 30$

4. $54\overline{)2{,}438}$

$2{,}500 \div 50 = 50$

$2{,}400 \div 60 = 40$

5. $68\overline{)4{,}831}$

$4{,}800 \div 60 = 80$

$4{,}900 \div 70 = 70$

6. $73\overline{)26{,}970}$

$24{,}000 \div 80 = 300$

$28{,}000 \div 70 = 400$

Divide by 2-Digit Divisors

A total of 6,501 people attended the local theater. A movie
was shown 20 times during a 5-day period. The same
number of people attended each showing except for the
first. How many people attended each showing?

Step 1

Decide where to place the
first digit in the quotient.
Are there enough thousands?
No, 6 < 20. Are there
enough hundreds? Yes,
65 > 20. The first digit
goes in the hundreds place.

$$20\overline{)6,501}$$

$$\square$$
$$20\overline{)6,501}$$

Step 2

Divide the hundreds. $20\overline{)65}$
Write the 3 in the hundreds
place.
Multiply. 20 × 3
Subtract. 65 − 60
Compare. 5 < 20

$$\begin{array}{r} 3 \\ 20\overline{)6,501} \\ -6\ 0 \\ \hline 5 \end{array}$$

Step 3

Divide the tens. $20\overline{)50}$
Write the 2 in the tens
place.
Multiply. 20 × 2
Subtract. 50 − 40
Compare. 10 < 20

$$\begin{array}{r} 32 \\ 20\overline{)6,501} \\ -6\ 0 \\ \hline 50 \\ -40 \\ \hline 10 \end{array}$$

Step 4

Divide the ones. $20\overline{)101}$
Write the 5 in the ones
place.
Multiply. 20 × 5
Subtract. 101 − 100
Compare. 1 < 20

$$\begin{array}{r} 325\ r1 \\ 20\overline{)6,501} \\ -6\ 0 \\ \hline 50 \\ -40 \\ \hline 101 \\ -100 \\ \hline 1 \end{array}$$

So, 325 people attended each showing of the movie, with 1 more person,
or 326 people, attending the first showing.

Follow the steps above to find each quotient.

1. $52\overline{)6,219}$ 119 r31

2. $81\overline{)9,017}$ 111 r26

3. $24\overline{)6,008}$ 250 r8

4. $17\overline{)92,418}$ 5,436 r6

5. $32\overline{)6,850}$ 214 r2

6. $41\overline{)87,409}$ 2,131 r38

Correcting Quotients

Maria collects postcards. She has 389 postcards in her collection. The cards are organized in albums that hold 48 postcards each. How many albums has Maria used?

Divide. $389 \div 48$

Step 1

Write two pairs of compatible numbers, and estimate the answer.

$$40\overline{)360} \quad \frac{9}{}$$

$$50\overline{)400} \quad \frac{8}{}$$

Step 2

Use one of your estimates.

The divisor, 48, is closer to 50. Use 8 as the first digit in the quotient.

Step 3

Divide.
Since $5 < 48$, the estimate is just right.

$$\begin{array}{r} 8 \text{ r}5 \\ 48\overline{)389} \\ -384 \\ \hline 5 \end{array}$$

So, Maria has 8 full albums and 1 album with only 5 postcards in it.

Use the steps above to find each quotient.

1. 3 r10
   $19\overline{)67}$

2. 3 r4
   $31\overline{)97}$

3. 4 r43
   $48\overline{)235}$

4. 13 r13
   $74\overline{)975}$

5. 8 r61
   $62\overline{)557}$

6. 10 r22
   $27\overline{)292}$

7. 9 r41
   $52\overline{)509}$

8. 9 r3
   $85\overline{)768}$

9. 71 r62
   $75\overline{)5,387}$

10. 170 r42
    $49\overline{)8,372}$

11. 642 r30
    $65\overline{)41,760}$

12. 1,102 r26
    $54\overline{)59,534}$

Practice Division

Ron's Record Shop received a shipment of 756 tapes. The
tapes were packaged in 28 cartons. Each carton held the
same number of tapes. How many tapes were in each carton?

Step 1 $756 \div 28$

$$28\overline{)756}^{\square}$$

Decide where to place the first digit.
Are there enough hundreds? No, $7 < 28$.
Place the first digit in the tens place.

Step 2

Divide the 75 tens.
Multiply. 28×2
Subtract. $75 - 56$
Compare. $19 < 28$

$$\begin{array}{r} 2 \\ 28\overline{)756} \\ -56 \\ \hline 19 \end{array}$$

Step 3

Divide the 196 ones.
Multiply. 28×7
Subtract. $196 - 196$

So, each carton held 27 tapes.

$$\begin{array}{r} 27 \\ 28\overline{)756} \\ -56 \\ \hline 196 \\ -196 \\ \hline 0 \end{array}$$

You can use multiplication to check the answer. Multiply the
divisor by the quotient. Add any remainder.

$28 \times 27 = 756$ The answer checks.

Follow the steps above to find each quotient. Check by multiplying.

1. $17\overline{)255}$ 15

2. $26\overline{)396}$ 15 r6

3. $33\overline{)458}$ 13 r29

4. $49\overline{)721}$ 14 r35

5. $45\overline{)6,004}$ 133 r19

6. $39\overline{)72,118}$ 1,849 r7

7. $15\overline{)497}$ 33 r2

8. $54\overline{)36,565}$ 677 r7

© Harcourt

Problem Solving Strategy: Predict and Test

Rhea has 253 stickers. She has them stored in equal groups in containers and has started a new container with 3 stickers in it. How many containers of stickers does she have? How many stickers are in each container?

Step 1

Subtract the 3 stickers in the new container from the 253 total number of stickers. $253 - 3 = 250$
There are 250 can be divided by 5.

Step 2

Use *predict and test* to find the number of equal groups in 250. The number ends with 0, so 250 can be divided by 5.

Step 3

Divide. $250 \div 5 = 50$

Check.
$$\begin{array}{r} 50 \\ \times\ 5 \\ \hline 250 \end{array} \qquad \begin{array}{r} 250 \\ +\ 3 \\ \hline 253 \end{array}$$
✔ The answer checks.

So, Rhea has 5 containers of stickers with 50 stickers in each container. There are 3 stickers in the new container.

Predict and test to solve.

1. James has 467 bookmarks in his collection. He has them stored in equal groups in boxes. He then starts a new box with 5 bookmarks in it. How many boxes of bookmarks does he have? How many bookmarks are in each box?
   Possible answer:
   6 boxes; 77 bookmarks

2. Nora baked 156 brownies. She is putting them into packages with an equal number of brownies in each. She eats 2 brownies. How many packages does she make? How many brownies are in each package?
   7 packages; 22 brownies

Algebra: Patterns in Decimal Division

Kara is dividing $3 equally into 5 boxes. How much money should go into each box?

$3 ÷ 5 = ?

Using a pattern can help you find the exact answer.

Write similar number sentences with zeros added to the dividends. The decimal point shifts one place to the left each time.

$$3{,}000 ÷ 5 = 600.0$$
$$300 ÷ 5 = 60.0$$
$$30 ÷ 5 = 6.0$$
$$3 ÷ 5 = 0.6$$

So, each box gets 0.6, or $0.60.

Complete each number sentence. Look for a pattern.

1. $3{,}000 ÷ 6 = 500$ **2.** $4{,}500 ÷ 5 = 900$ **3.** $6{,}400 ÷ 8 = 800$ **4.** $2{,}800 ÷ 7 = 400$

$\underline{300} ÷ 6 = 50$ $\underline{450} ÷ 5 = 90$ $\underline{640} ÷ 8 = 80$ $280 ÷ 7 = \underline{40}$

$30 ÷ 6 = \underline{5}$ $45 ÷ 5 = \underline{9}$ $64 ÷ 8 = \underline{8}$ $28 ÷ 7 = \underline{4}$

$3 ÷ 6 = \underline{0.5}$ $4.5 ÷ 5 = \underline{0.9}$ $6.4 ÷ 8 = \underline{0.8}$ $2.8 ÷ 7 = \underline{0.4}$

Use a pattern to write the quotients.

5. $400 ÷ 8 = \underline{50}$ **6.** $600 ÷ 4 = \underline{150}$ **7.** $800 ÷ 5 = \underline{160}$

$40 ÷ 8 = \underline{5}$ $60 ÷ 4 = \underline{15}$ $80 ÷ 5 = \underline{16}$

$4 ÷ 8 = \underline{0.5}$ $6 ÷ 4 = \underline{1.5}$ $8 ÷ 5 = \underline{1.6}$

8. $1{,}400 ÷ 7 = \underline{200}$ **9.** $13{,}000 ÷ 5 = \underline{2{,}600}$ **10.** $2{,}700 ÷ 9 = \underline{300}$

$140 ÷ 7 = \underline{20}$ $1{,}300 ÷ 5 = \underline{260}$ $270 ÷ 9 = \underline{30}$

$14 ÷ 7 = \underline{2}$ $130 ÷ 5 = \underline{26}$ $27 ÷ 9 = \underline{3}$

$1.4 ÷ 7 = \underline{0.2}$ $13 ÷ 5 = \underline{2.6}$ $2.7 ÷ 9 = \underline{0.3}$

Decimal Division

You can use a centimeter ruler to help you divide 1.4 by 2.

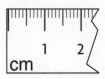

1 stands for 1 centimeter.

Each space stands for $\frac{1}{10}$, or 0.1, cm.

Step 1

Find 1.4 centimeters on the ruler. Count the number of spaces.

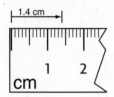

There are 14 spaces.

Step 2

Divide the number of spaces by 2.
$14 \div 2 = 7$

Count over 7 spaces.

The seventh space is 0.7 cm.
So $1.4 \div 2 = 0.7$.

Use the centimeter ruler to find the quotient.

1. $2.4 \div 6 = \underline{0.4}$

2. $2.5 \div 5 = \underline{0.5}$

3. $1.8 \div 3 = \underline{0.6}$

4. $2.4 \div 4 = \underline{0.6}$

5. $2.1 \div 7 = \underline{0.3}$

6. $1.6 \div 8 = \underline{0.2}$

7. $3.9 \div 3 = \underline{1.3}$

8. $3.3 \div 3 = \underline{1.1}$

9. $0.8 \div 8 = \underline{0.1}$

© Harcourt

Divide Decimals by Whole Numbers

You can use a centimeter ruler to help you divide 3.6 by 2.

1 stands for 1 centimeter.
Each space stands for $\frac{1}{10}$ or 0.1 cm.

Step 1

Find 3.6 centimeters on the ruler.

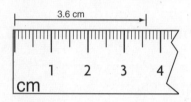

Step 2

There are 3 whole centimeters. Divide them into 2 equal groups. There is 1 centimeter in each group with 1.6 centimeters left over.

$$\begin{array}{r} 1 \\ 2\overline{)3.6} \\ -2.0 \\ \hline 1.6 \end{array}$$

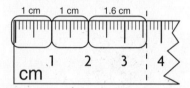

Step 3

Count the spaces for the remaining 1.6 cm. There are 16 spaces. Divide them into 2 groups. There are 2 groups of 8 spaces. Each group is 0.8 centimeter.

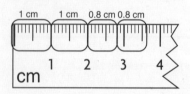

Step 4

There are 2 equal groups of 1.8 centimeters.

So, 3.6 ÷ 2 = 1.8

$$\begin{array}{r} 1.8 \\ 2\overline{)3.6} \\ -2.0 \\ \hline 1.6 \\ -1.6 \\ \hline 0 \end{array}$$

Use the ruler to find the quotient.

1. $3\overline{)3.6}$ 1.2

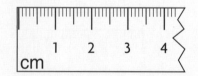

2. $2\overline{)2.8}$ 1.4

3. $2\overline{)3.2}$ 1.6

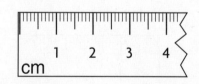

4. $4\overline{)4.0}$ 1.0

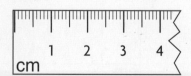

5. $2\overline{)1.2}$ 0.6

6. $4\overline{)3.6}$ 0.9

Problem Solving Strategy

Compare Strategies

Problem Since the school year began, Jill has grown 0.75 inches. Now she measures 58.5 inches. What did she measure when the year began?

What strategy can you use?

Work backward: What information can you use to find out how tall Jill was when the year began? You can start by using the information at the end. Now she is 58.5 inches. Then use the fact that she grew 0.75 inch. Work backward to find out how tall she was at the beginnning of the year.

58.5 = current height

0.75 = height she grew   58.5 − 0.75 = 57.75

57.75 inches = height at beginning of school year.

Work backward

You can also use *predict and test.*

Predict: She was 57 inches.   Test: 57 + 0.75 = 57.75; too low

Predict again, using a higher number.

Predict: She was 57.75 inches.

Predict and Test

Test: 57.75 + 0.75 = 58.5 inches

Solve and write the problem solving strategy you used: *work backward* or *predict and test.*

1. Anthony started with his favorite number. Then he subtracted 7 from it. He multiplied this difference by 3 and then added 5. Finally he divided this number by 11. His end result was 1. What was Anthony's favorite number?

   _____9; work backward_____

2. Forty-seven baseball players need a ride to the play-off game. Each car has seat belts for 4 players and can make 2 trips. How many cars will be needed?

   _____6 cars; predict and test_____

3. The sum of 2 numbers is 40 and their difference is 2. What are the two numbers?

   _____19 and 21; predict and test_____

4. The school spent $438.75 to buy art supplies and gym supplies. The total cost of the art supplies was $230.60. How much was spent on the gym supplies?

   _____$208.15; work backward_____

Divide to Change a Fraction to a Decimal

Fractions can be written as decimals by dividing the numerator by the denominator.

$$\frac{\text{numerator}}{\text{denominator}} \rightarrow \text{denominator}\overline{)\text{numerator}}$$

To write $\frac{3}{5}$ as a decimal, divide 3 by 5.

$$\frac{3}{5} = \begin{array}{r} 0.6 \\ 5\overline{)3.0} \\ \underline{3.0} \\ 0 \end{array} \leftarrow \text{numerator}$$

denominator ↑

Write as a decimal.

1. $\frac{3}{50}$ 0.06

2. $\frac{4}{10}$ 0.4

3. $\frac{16}{100}$ 0.16

4. $\frac{3}{4}$ 0.75

5. $\frac{20}{40}$ 0.5

6. $\frac{8}{10}$ 0.8

7. $\frac{15}{20}$ 0.75

8. $\frac{4}{5}$ 0.8

9. $\frac{42}{50}$ 0.84

10. $\frac{10}{25}$ 0.4

11. $\frac{63}{100}$ 0.63

12. $\frac{1}{8}$ 0.125

13. $\frac{1}{4}$ 0.25

14. $\frac{4}{8}$ 0.5

15. $\frac{6}{25}$ 0.24

16. $\frac{3}{8}$ 0.375

17. $\frac{3}{15}$ 0.2

18. $\frac{721}{1,000}$ 0.721

19. $\frac{4}{100}$ 0.04

20. $\frac{5}{8}$ 0.625

21. $\frac{14}{25}$ 0.56

22. $\frac{47}{50}$ 0.94

23. $\frac{8}{1,000}$ 0.008

24. $\frac{7}{8}$ 0.875

25. $\frac{30}{1,000}$ 0.030

Algebra: Patterns in Decimal Division

Helen has $2.00, and she is putting $0.50 into each box.
How many boxes can she fill?

$2.00 \div 0.50 = \underline{?}$

Using a pattern can help you find the exact answer.

$$20 \div 5 = 4$$

Notice that moving the decimal point one place
to the left for the divisor and the dividend will
give the same answer each time.

$$2.0 \div 0.5 = 4$$

$$0.2 \div 0.05 = 4$$

So, She can fill 4 boxes with $0.50 in each box.

Complete each number sentence. Look for patterns.

1. $240 \div 6 = 40$ **2.** $320 \div 8 = 40$ **3.** $490 \div 7 = 70$ **4.** $540 \div 9 = 60$

$24.0 \div \underline{0.6} = 40$ $32.0 \div \underline{0.8} = 40$ $49.0 \div \underline{0.7} = 70$ $54.0 \div \underline{0.9} = 60$

$2.4 \div 0.06 = \underline{40}$ $3.2 \div 0.08 = \underline{40}$ $4.9 \div 0.07 = \underline{70}$ $5.4 \div 0.09 = \underline{60}$

Use patterns to write the quotients.

5. $36 \div 6 = \underline{6}$ **6.** $42 \div 7 = \underline{6}$ **7.** $54 \div 6 = \underline{9}$

$3.6 \div 0.6 = \underline{6}$ $4.2 \div 0.7 = \underline{6}$ $5.4 \div 0.6 = \underline{9}$

$0.36 \div 0.06 = \underline{6}$ $0.42 \div 0.07 = \underline{6}$ $0.54 \div 0.06 = \underline{9}$

8. $32 \div 4 = \underline{8}$ **9.** $56 \div 8 = \underline{7}$ **10.** $63 \div 7 = \underline{9}$

$3.2 \div 0.4 = \underline{8}$ $5.6 \div 0.8 = \underline{7}$ $6.3 \div 0.7 = \underline{9}$

$0.32 \div 0.04 = \underline{8}$ $0.56 \div 0.08 = \underline{7}$ $0.63 \div 0.07 = \underline{9}$

11. $28 \div 4 = \underline{7}$ **12.** $24 \div 6 = \underline{4}$ **13.** $48 \div 6 = \underline{8}$

$2.8 \div 0.4 = \underline{7}$ $2.4 \div 0.6 = \underline{4}$ $4.8 \div 0.6 = \underline{8}$

$0.28 \div 0.04 = \underline{7}$ $0.24 \div 0.06 = \underline{4}$ $0.48 \div 0.06 = \underline{8}$

Divide with Decimals

Use the following rules to divide a decimal by another decimal:

1.) Move the <u>decimal point in the divisor</u> to the **right** to change it to a whole number.

2.) Move the <u>decimal point in the dividend</u> to the **right** the same number of places as you did in the divisor.

3.) Put the <u>decimal point in the quotient</u> directly above the new decimal point in the dividend.

4.) Divide the numbers to obtain the quotient.

Example:

Move the <u>decimal point</u> 2 places to the **right.**

Divide the numbers to obtain the quotient.

$$\text{quotient} = \mathbf{6}.$$
$$0.07\overline{)0.42} = 7.\overline{)042}.$$

Find the quotient. Check by multiplying.

1. $3.2 \div 0.4 =$ ___8___

   Check: $0.4 \times 8 = 3.2$

2. $6.4 \div 0.8 =$ ___8___

   Check: $0.8 \times 8 = 6.4$

3. $0.21 \div 0.07 =$ ___3___

   Check: $0.07 \times 3 = 0.21$

4. $0.50 \div 0.25 =$ ___2___

   Check: $0.25 \times 2 = 0.50$

5. $3.9 \div 1.3 =$ ___3___

   Check: $1.3 \times 3 = 3.9$

6. $0.96 \div 0.24 =$ ___4___

   Check: $0.24 \times 4 = .96$

7. $2.4 \div 0.4 =$ ___6___

   Check: $0.4 \times 6 = 2.4$

8. $0.49 \div 0.07 =$ ___7___

   Check: $0.07 \times 7 = 0.49$

Decimal Division

When dividing a decimal by another decimal, you must change the divisor to a whole number by multiplying the divisor by 10 or 100. Whatever number you use to multiply the divisor, you must also use to multiply the dividend.

Divide 52.36 by 0.28.

Write the division problem on graph paper.

Move the decimal point in the divisor by multiplying by 100. 0.28 changes to 28.0.

Move the decimal point in the dividend by multiplying it by 100. 52.36 changes to 5236.0.

Divide as if you were working with whole numbers.

$$
\begin{array}{r}
187. \\
0.28\overline{)5236.} \\
-28 \\
\hline
243 \\
-224 \\
\hline
196 \\
-196 \\
\hline
000
\end{array}
$$

Find the quotient. Check by multiplying.

1. $0.5\overline{)9.25}$ quotient: 18.5   Check:
18.5
× 0.5
9.25

2. $5.2\overline{)4.524}$ quotient: 0.87   Check:
0.87
× 5.2
4.524

3. $0.8\overline{)0.896}$ quotient: 1.12   Check:
1.12
× 0.8
0.896

4. $0.56\overline{)24.472}$ quotient: 43.7   Check:
43.7
× 0.56
24.472

5. $24.78 \div 0.3 = \underline{82.6}$ Check:
82.6
× 0.3
24.78

6. $\$39.00 \div \$0.52 = \underline{75}$ Check:
$75
× 0.52
$39.00

7. $9.144 \div 0.36 = \underline{25.4}$ Check:
25.4
× 0.36
9.144

8. $\$1.84 \div \$0.04 = \underline{46}$ Check:
$0.04
× 46
$1.84

Problem Solving Skill

Choose the Operation

Ricardo and his two friends raise small animals. Ricardo buys rabbits, hamsters, mice, and gerbils. If Ricardo and his friends each take an equal number of animals, how many animals will each person get?

Type of Animal	Number
Rabbits	15
Hamsters	27
Mice	36
Gerbils	9

There are 15 rabbits for 3 people.

Should you multiply?   or   Should you divide?

$15 \times 3 = 45$ $15 \div 3 = 5$

Which answer makes more sense? Since they bought only 15 rabbits, 5 rabbits each makes the most sense. You should divide.

For Problems 1–6, use the table to solve each problem. Name the operation you used.

Type of Food	Amount (in pounds)
Rabbit Food	186.3
Hamster Food	53.1
Mouse Food	26.9
Gerbil Food	12.6

1. Ricardo and his two friends purchase animal food. They share what they buy equally. What is Ricardo's share of the rabbit food?

   division;
   $186.3 \div 3 = 62.1$ lb

2. Ricardo buys the same amount of gerbil food each month for 5 months. How much gerbil food does Ricardo buy?

   multiplication;
   $12.6 \times 5 = 63$ lb

3. Ricardo pays $1.25 per pound for a month's worth of gerbil food. How much does the gerbil food cost in all?

   multiplication;
   $12.6 \times \$1.25 = \$15.75$

4. Ricardo spent $37.17 buying hamster food. What was the cost per pound for the hamster food?

   division;
   $\$37.17 \div 53.1 = \$0.70$ lb

5. How much animal food do Ricardo and his friends buy in all?

   addition;
   $186.3 + 53.1 + 26.9 + 12.6 = 278.9$ lb

6. What is Ricardo's share of the hamster food?

   division;
   $53.1 \div 3 = 17.7$ lb

Divisibility

The rules for divisibility by 3 and 9 are special. They depend on finding the sum of the digits.

- A number is divisible by 3 if the sum of the digits of the number is divisible by 3.

- A number is divisible by 9 if the sum of the digits of the number is divisible by 9.

---

1. Decide if 615 is divisible by 3.

   a. What is the sum of the digits 6, 1, and 5? _____12_____

   b. Is 12 divisible by 3? _____yes_____

   c. Is 615 divisible by 3? _____yes_____

2. Decide if 615 is divisible by 9.

   a. What is the sum of the digits 6, 1, and 5? _____12_____

   b. Is 12 divisible by 9? _____no_____

   c. Is 615 divisible by 9? _____no_____

Tell if each number is divisible by 3 or 9. Write 3, 9, or neither.

3. 90	4. 315	5. 390	6. 405
3, 9	3, 9	3	3, 9

7. 75	8. 4,770	9. 320	10. 3,705
3	3, 9	neither	3

11. 801	12. 408	13. 117	14. 490
3, 9	3	3, 9	neither

15. 81	16. 906	17. 432	18. 235
3, 9	3	3, 9	neither

19. 123	20. 684	21. 963	22. 91
3	3, 9	3, 9	neither

Multiples and Least Common Multiples

Sam and Mary love to count. Sam counts by 3's and Mary counts by 4's.

           Sam   Mary

3, 6, 9, (12), 15, 18, 21, (24), 27, 30, 33, . . .   4, 8, (12), 16, 20, (24), 28, 32, 36, 40, . . .

Sam and Mary both say the numbers 12 and 24. These numbers are called the **common multiples** of 3 and 4. The first common multiple is 12, so it is called the **least common multiple** of 3 and 4.

List the first 6 multiples of the number.

1. 2   **2.** 5   **3.** 6

  2, 4, 6, 8, 10, 12   5, 10, 15, 20, 25, 30   6, 12, 18, 24, 30, 36

4. 7   **5.** 8   **6.** 9

  7, 14, 21, 28, 35, 42   8, 16, 24, 32, 40, 48   9, 18, 27, 36, 45, 54

7. 10   **8.** 11   **9.** 12

  10, 20, 30, 40, 50,   11, 22, 33, 44, 55,   12, 24, 36, 48, 60,

  60   66   72

Find the first 2 common multiples of each pair of numbers.

10. 2 and 5   **11.** 4 and 8   **12.** 6 and 8   **13.** 4 and 12

  10, 20   8, 16   24, 48   12, 24

Find the least common multiple of each pair of numbers.

14. 3 and 8   **15.** 6 and 9   **16.** 5 and 8   **17.** 3 and 7

  24   18   40   21

Greatest Common Factor

You can find the **greatest common factor** of two numbers. It is the greatest factor that the two numbers have in common.

Find the greatest common factor of 9 and 15.

Step 1	**Step 2**	**Step 3**
List all the factors of each number.	Note the common factors.	Which factor is greater?
9: 1, 3, 9	The common factors of 9 and 15 are 1 and 3.	3 is greater than 1.
15: 1, 3, 5, 15		

So, the greatest common factor of 9 and 15 is 3.

Use the factors given to find the greatest common factor (GCF) for each pair of numbers.

1. 10: 1, 2, 5, 10
   25: 1, 5, 25

   GCF _____5_____

2. 18: 1, 2, 3, 6, 9, 18
   21: 1, 3, 7, 21

   GCF _____3_____

3. 28: 1, 2, 4, 7, 14, 28
   35: 1, 5, 7, 35

   GCF _____7_____

4. 21: 1, 3, 7, 21
   49: 1, 7, 49

   GCF _____7_____

List the factors of each number. Write the greatest common factor (GCF) for each pair of numbers. The first one is done for you.

5. 8 _____1, 2, 4, 8_____

   12 _____1, 2, 3, 4, 6, 12_____

   GCF _____4_____

6. 6 _____1, 2, 3, 6_____

   24 _____1, 2, 3, 4, 6, 8, 12, 24_____

   GCF _____6_____

7. 9 _____1, 3, 9_____

   27 _____1, 3, 9, 27_____

   GCF _____9_____

8. 4 _____1, 2, 4_____

   14 _____1, 2, 7, 14_____

   GCF _____2_____

Problem Solving Skill

Identify Relationships

Identifying relationships can help you solve some word problems.

There is a relationship between the product of two numbers and the product of their least common multiple (LCM) and greatest common factor (GCF).

Example:

• Find the relationship between the product of 6 and 9, and the product of their LCM and GCF.

   The LCM of 6 and 9 is 18.
   The GCF of 6 and 9 is 3.

   $6 \times 9 = \textbf{54}$ $\text{LCM} \times \text{GCF} = 18 \times 3 = \textbf{54}$

So, the product of two numbers is equal to the product of their LCM and GCF.

Use the relationship between the given numbers to complete the table.

First Number	Second Number	Product of Numbers	Product of LCM and GCF
6	15	90	90
8	4	32	32
7	8	56	56
12	3	36	36

Use the relationships between the given numbers to solve.

1. The product of the LCM and GCF of 4 and another number is 36. What is the other number?

   9

2. The product of two numbers is 98. The GCF of the two numbers is 7. What is their LCM?

   14

3. The product of the LCM and GCF of two numbers is 55, and neither of the numbers is 1. What are the two numbers?

   5 and 11

4. The product of two numbers is 320. The GCF of the two numbers is 4, and one of the numbers is 16. What is the other number?

   20

Name _____

Prime and Composite Numbers

You can use squares to see if a number is prime or composite.

A prime number has exactly two factors, 1 and the number itself.

Is the number 5 prime or composite?

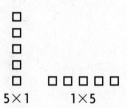

5×1   1×5

Only 2 arrangements of squares are possible (5 × 1, 1 × 5). The number 5 has exactly two factors, so it is a prime number.

A composite number has more than two factors.

Is the number 8 prime or composite?

8×1   1×8   4×2   2×4

More than 2 arrangements of squares are possible (8 × 1, 1 × 8, 4 × 2, 2 × 4). The number 8 has more than two factors, so it is a composite number.

Draw squares to see if each number is prime or composite. Write *prime* or *composite*. Check students' drawings.

1. 7 _____prime_____

2. 6 _____composite_____

Write the possible arrangements of squares for each number. Then write *prime* or *composite*. The first one is done for you.

3. 4 _____ 1 × 4, 4 × 1, 2 × 2; composite _____

4. 9 _____ 1 × 9, 9 × 1, 3 × 3; composite _____

5. 10 _____ 1 × 10, 10 × 1, 2 × 5, 5 × 2; composite _____

6. 11 _____ 1 × 11, 11 × 1; prime _____

7. 12 _____ 1 × 12, 12 × 1, 3 × 4, 4 × 3, 6 × 2, 2 × 6; composite _____

8. 13 _____ 1 × 13, 13 × 1; prime _____

© Harcourt

Introduction to Exponents

Exponents are also called "powers."

$10 \times 10 = 10^2$
10^2 = the second power of ten, or ten squared

$10 \times 10 \times 10 = 10^3$

10^3 = the third power of ten, or ten cubed

Show the eighth power of ten in four different ways.

Exponent Form	Expanded Form	Standard Form	Word Form
10^8	$10^8 = 10 \times 10 \times$ $10 \times 10 \times 10 \times$ $10 \times 10 \times 10$	100,000,000	One hundred million

---

Write in expanded form.

1. 100

$\underline{10 \times 10}$
_____

2. 10,000

$\underline{10 \times 10 \times 10}$
$\underline{\times 10}$

3. 1,000

$\underline{10 \times 10 \times 10}$
_____

4. 100,000

$\underline{10 \times 10 \times 10}$
$\underline{\times 10 \times 10}$

5. 10^6

$\underline{10 \times 10 \times 10}$
$\underline{\times 10 \times 10 \times 10}$

6. 10^3

$\underline{10 \times 10 \times 10}$
_____

7. 10^5

$\underline{10 \times 10 \times 10}$
$\underline{\times 10 \times 10}$

8. 10^7

$\underline{10 \times 10 \times 10 \times 10}$
$\underline{\times 10 \times 10 \times 10}$

Write in exponent form.

9. 100,000

$\underline{10^5}$

10. 10,000

$\underline{10^4}$

11. 10,000,000

$\underline{10^7}$

12. 1,000

$\underline{10^3}$

13. the third power of ten

$\underline{10^3}$

14. the seventh power of ten

$\underline{10^7}$

15. the ninth power of ten

$\underline{10^9}$

16. the tenth power of ten

$\underline{10^{10}}$

Evaluate Expressions with Exponents

What number does 6^5 represent?

6 is called the **base.** The 5 is called the **exponent.** The exponent tells you how many times the base is used as a factor.

$6^5 = 6 \times 6 \times 6 \times 6 \times 6 = 7{,}776$

$4^3 = 4 \times 4 \times 4 = 64$

4 is the base and 3 is the exponent.

Write the base and the exponent.

1. 4^6 | **2.** 6^4 | **3.** 9^{18} | **4.** 5^7

Base | Base | Base | Base

_____4_____ | _____6_____ | _____9_____ | _____5_____

Exponent | Exponent | Exponent | Exponent

_____6_____ | _____4_____ | _____18_____ | _____7_____

Write the equal factors.

5. 9^9 | **6.** 6^7 | **7.** 3^9 | **8.** 12^6

$9 \times 9 \times 9 \times 9$ | $6 \times 6 \times 6 \times 6$ | $3 \times 3 \times 3 \times 3$ | $12 \times 12 \times 12$

$\times 9 \times 9 \times 9$ | $\times 6 \times 6 \times 6$ | $\times 3 \times 3 \times 3$ | $\times 12 \times 12 \times 12$

$\times 9 \times 9$ | | $\times 3 \times 3$ |

9. 14^2 | **10.** 8^{10} | **11.** 11^{11} | **12.** 24^4

14×14 | $8 \times 8 \times 8 \times 8$ | $11 \times 11 \times 11 \times 11$ | $24 \times 24 \times 24$

| $\times 8 \times 8 \times 8$ | $\times 11 \times 11 \times 11 \times 11$ | $\times 24$

| $\times 8 \times 8 \times 8$ | $\times 11 \times 11 \times 11$ |

Find the value.

13. 4^6 | **14.** 6^4 | **15.** 9^3 | **16.** 5^2

_____4,096_____ | _____1,296_____ | _____729_____ | _____25_____

Name _____

Exponents and Prime Factors

You can think about prime factorization as a series of division problems.

Begin with the number you need to factor: **48**

What is the least possible prime number that divides 48? **2**

Keep dividing by prime divisors until you get 1 as a quotient.

Divide 2 into 48.

1. Is the quotient 1? No. $\quad\quad 2\overline{)48} \;\; (24)$

Repeat the process.

2. Is the quotient 1? No. $\quad\quad 2\overline{)24} \;\; (12)$

Repeat the process.

3. Is the quotient 1? No. $\quad\quad 2\overline{)12} \;\; (6)$

Repeat the process.

4. Is the quotient 1? No. $\quad\quad 2\overline{)6} \;\; (3)$

Repeat the process.

5. Is the quotient 1? Yes. $\quad\quad 3\overline{)3} \;\; (1)$

Stop.

Write the prime divisors as factors of 48.

$48 = 2 \times 2 \times 2 \times 2 \times 3$

Use what you know about exponents to write the factors.

$48 = 2^4 \times 3$

Write the prime factorization of the number. Use exponents when possible.

1. 12

$\underline{\quad 2^2 \times 3 \quad}$

2. 24

$\underline{\quad 2^3 \times 3 \quad}$

3. 28

$\underline{\quad 2^2 \times 7 \quad}$

4. 45

$\underline{\quad 3^2 \times 5 \quad}$

5. 36

$\underline{\quad 2^2 \times 3^2 \quad}$

6. 125

$\underline{\quad 5^3 \quad}$

7. 256

$\underline{\quad 2^8 \quad}$

8. 81

$\underline{\quad 3^4 \quad}$

Name _____

Relate Decimals to Fractions

You can write a fraction or a decimal to tell what part is shaded.

Model	Fraction	Decimal			Read
	$\dfrac{4 \text{ shaded parts}}{100 \text{ parts}}$	O	T	H	four hundredths
		0 .	0	4	
	$\dfrac{25 \text{ shaded parts}}{100 \text{ parts}}$	O	T	H	twenty-five hundredths
		0 .	2	5	

Complete the table.

	Model	Fraction	Decimal			Read
1.		$\dfrac{\boxed{76} \text{ shaded parts}}{\boxed{100} \text{ parts}}$	O	T	H	seventy-six hundredths
			0 .	7	6	
2.		12 shaded parts / 100 parts	O	T	H	twelve hundredths
			0 .	1	2	
3.		22 shaded parts / 100 parts	O	T	H	twenty-two hundredths
			0 .	2	2	
4.		5 shaded parts / 10 parts	O	T	H	five tenths
			0 .	5	0	
5.		38 shaded parts / 100 parts	O	T	H	thirty-eight hundredths
			0 .	3	8	

Name _____

Equivalent Fractions

You can use different fractions to name the same amount.

Fractions that name the same amount are called **equivalent fractions**.

You can find equivalent fractions in three ways.

Use a number line.	Multiply both the numerator and the denominator by the same number.	Divide both the numerator and the denominator by the same number.
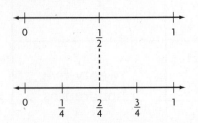 You can see that $\frac{1}{2} = \frac{2}{4}$, so they are equivalent fractions.	$\frac{1}{3} = \frac{1 \times 3}{3 \times 3} = \frac{3}{9}$ The fraction $\frac{1}{3}$ names the same amount as $\frac{3}{9}$, so they are equivalent fractions.	$\frac{6}{8} = \frac{6 \div 2}{8 \div 2} = \frac{3}{4}$ The fractions $\frac{6}{8}$ and $\frac{3}{4}$ are equal, so they are equivalent fractions.

Use the number lines to find out if the fractions are equivalent.
Write *yes* or *no*.

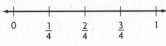

1. $\frac{1}{4} = \frac{3}{12}$ _yes_

2. $\frac{8}{12} = \frac{3}{4}$ _no_

Multiply both the numerator and the denominator to name an equivalent fraction.

3. $\frac{3}{8} = \frac{3 \times 2}{8 \times 2} = \frac{6}{16}$

4. $\frac{2}{3} = \frac{2 \times 5}{3 \times 5} = \frac{10}{15}$

5. $\frac{1}{7} = \frac{1 \times 4}{7 \times 4} = \frac{4}{28}$

6. $\frac{4}{5} = \frac{4 \times 3}{5 \times 3} = \frac{12}{15}$

Divide both the numerator and the denominator to name an equivalent fraction.

7. $\frac{12}{16} = \frac{12 \div 4}{16 \div 4} = \frac{3}{4}$

8. $\frac{7}{28} = \frac{7 \div 7}{28 \div 7} = \frac{1}{4}$

9. $\frac{10}{15} = \frac{10 \div 5}{15 \div 5} = \frac{2}{3}$

10. $\frac{16}{24} = \frac{16 \div 8}{24 \div 8} = \frac{2}{3}$

© Harcourt

Compare and Order Fractions

The three fractions $\frac{2}{3}$, $\frac{3}{4}$, and $\frac{2}{6}$ are arguing about who is the greatest.

You can settle the argument by finding a common multiple for the denominators.

Step 1	Step 2	Step 3
Find the product of all three denominators.	Rename each fraction so that 72 is the denominator.	Compare the numerators. Put them in order from least to greatest.

Step 1

Find the product of all three denominators.

$3 \times 4 \times 6 = 72$

72 is a common multiple.

Use it for the denominator.

Step 2

Rename each fraction so that 72 is the denominator.

$$\frac{2 \times 24}{3 \times 24} = \frac{48}{72}$$

$$\frac{3 \times 18}{4 \times 18} = \frac{54}{72}$$

$$\frac{2 \times 12}{6 \times 12} = \frac{24}{72}$$

Step 3

Compare the numerators. Put them in order from least to greatest.

$$\frac{24}{72} < \frac{48}{72} < \frac{54}{72}$$

$$\frac{2}{6} < \frac{2}{3} < \frac{3}{4}$$

So, $\frac{3}{4}$ is the greatest fraction.

Find the product of the denominators.

1. $\frac{2}{5}, \frac{3}{4}, \frac{5}{7}$

   _____ 140 _____

2. $\frac{2}{9}, \frac{1}{3}, \frac{1}{2}$

   _____ 54 _____

3. $\frac{1}{2}, \frac{1}{5}, \frac{1}{8}$

   _____ 80 _____

Rename the fractions by using a common denominator.

4. $\frac{2}{5}, \frac{3}{4}, \frac{5}{7}$

   $\frac{56}{140}, \frac{105}{140}, \frac{100}{140}$

5. $\frac{2}{9}, \frac{1}{3}, \frac{1}{2}$

   $\frac{12}{54}, \frac{18}{54}, \frac{27}{54}$

6. $\frac{1}{2}, \frac{1}{5}, \frac{1}{8}$

   $\frac{40}{80}, \frac{16}{80}, \frac{10}{80}$

Compare and order from least to greatest.

7. $\frac{2}{5}, \frac{3}{4}, \frac{5}{7}$

   $\frac{2}{5}, \frac{5}{7}, \frac{3}{4}$

8. $\frac{2}{9}, \frac{1}{3}, \frac{1}{2}$

   $\frac{2}{9}, \frac{1}{3}, \frac{1}{2}$

9. $\frac{1}{2}, \frac{1}{5}, \frac{1}{8}$

   $\frac{1}{8}, \frac{1}{5}, \frac{1}{2}$

Simplest Form

You can use fraction bars to find the simplest form of a fraction.

Find the simplest form for $\frac{3}{12}$.

Step 1   Model $\frac{3}{12}$ with fraction bars.

| $\frac{1}{12}$ | $\frac{1}{12}$ | $\frac{1}{12}$ | $\frac{3}{12}$ |

Step 2   Line up other fraction bars to find all the equivalent fractions for $\frac{3}{12}$. You can see that $\frac{2}{8}$ and $\frac{1}{4}$ are equivalent fractions for $\frac{3}{12}$.

| $\frac{1}{12}$ | $\frac{1}{12}$ | $\frac{1}{12}$ | $\frac{3}{12}$ |

| $\frac{1}{8}$ | $\frac{1}{8}$ | $\frac{2}{8}$ |

| $\frac{1}{4}$ | $\frac{1}{4}$ |

Step 3   The equivalent fraction that has the largest fraction bar possible is in the simplest form.

So, $\frac{1}{4}$ is the simplest form of $\frac{3}{12}$.

Use the fraction bar outlines below to model each fraction and equivalent fractions. Divide the outline into equal parts or keep it whole. Write the fraction in its simplest form. Check students' models.

1. $\frac{9}{12}$ | $\frac{1}{12}$ | $\frac{1}{12}$ | $\frac{1}{12}$ | $\frac{1}{12}$ | $\frac{1}{12}$ | $\frac{1}{12}$ | $\frac{1}{12}$ | $\frac{1}{12}$ | $\frac{1}{12}$ | $\frac{9}{12}$ _____

| $\frac{1}{8}$ | $\frac{1}{8}$ | $\frac{1}{8}$ | $\frac{1}{8}$ | $\frac{1}{8}$ | $\frac{1}{8}$ | $\frac{6}{8}$ _____ Simplest form $\frac{3}{4}$ _____

| $\frac{1}{4}$ | $\frac{1}{4}$ | $\frac{1}{4}$ | $\frac{3}{4}$ _____

2. $\frac{4}{12}$ | $\frac{1}{12}$ | $\frac{1}{12}$ | $\frac{1}{12}$ | $\frac{1}{12}$ | $\frac{4}{12}$ _____

| $\frac{1}{6}$ | $\frac{1}{6}$ | $\frac{2}{6}$ _____ Simplest form $\frac{1}{3}$ _____

| $\frac{1}{3}$ | $\frac{1}{3}$ _____

Name _____

Understand Mixed Numbers

John drank $2\frac{3}{4}$ cartons of milk with his lunch.

The number $2\frac{3}{4}$ is a mixed number. A **mixed number** is made up of a whole number and a fraction.

In the mixed number $2\frac{3}{4}$, the whole number 2 represents two whole cartons of milk.

In the mixed number $2\frac{3}{4}$, the fraction $\frac{3}{4}$ represents a part of another carton.

You can divide all three cartons into 4 equal parts to show how many fourths John drank.

There are 11 shaded parts. Each part is $\frac{1}{4}$ carton.

So, John drank $\frac{11}{4}$, or $2\frac{3}{4}$, cartons of milk.

$$2\frac{3}{4}$$

Whole Number ↗   ↖ Fraction

2 cartons $\frac{3}{4}$ carton

 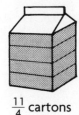 

$\frac{11}{4}$ cartons

Write both a fraction and a mixed number for each figure.

1.

$\frac{18}{5}$, $3\frac{3}{5}$

2.

$\frac{21}{5}$, $4\frac{1}{5}$

3.

$\frac{17}{6}$, $2\frac{5}{6}$

4.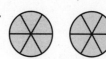

$\frac{12}{6}$, 2

5.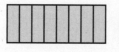

$\frac{11}{8}$, $1\frac{3}{8}$

6.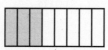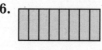

$\frac{23}{8}$, $2\frac{7}{8}$

© Harcourt

Problem Solving Strategy

Make a Model

Trisha spent $\frac{3}{4}$ hour on math homework, $\frac{3}{8}$ hour on science, and $\frac{1}{2}$ hour on language arts. Which homework did she spend the most time on?

You can *make a model* to solve this problem.

Step 1 For each fraction, draw a box. Shade the box to show the fraction.

Step 2 Find the LCD, and divide each box into that many equal parts.

The LCD is 8.

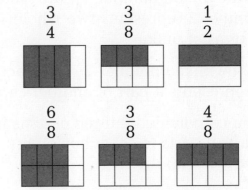

$\frac{3}{4}$ $\frac{3}{8}$ $\frac{1}{2}$

$\frac{6}{8}$ $\frac{3}{8}$ $\frac{4}{8}$

Step 3 Compare the numerators. $\frac{6}{8}$ is the greatest fraction.

So, Trisha spent the most time on math homework.

Make a model to solve.

1. Joe loves to cook. Last weekend he used flour in three different recipes. The amounts were $\frac{3}{4}$ cup, $\frac{2}{4}$ cup, and $\frac{1}{4}$ cup. What was the least amount called for?

   $\frac{1}{4}$ cup of flour

2. Karen walked $\frac{5}{6}$ of a mile from her house to a friend's house. Joe walked $\frac{7}{12}$ of a mile to his friend's house. Who walked a greater distance?

   Karen

3. Nick bought $\frac{2}{3}$ pound ground beef, $\frac{11}{12}$ pound ground turkey, and $\frac{3}{4}$ pound ground veal. Which meat did he buy the most of?

   turkey

4. In the store display, $\frac{2}{5}$ of the T-shirts were yellow and $\frac{1}{4}$ were blue. Were there more yellow or blue T-shirts?

   yellow

Add and Subtract Like Fractions

The denominators must be the same when adding or
subtracting fractions.

Add $\frac{2}{6} + \frac{1}{6}$.

Step 1	**Step 2**	**Step 3**
Are the denominators the same? Yes.	Add the numerators. The denominator stays the same.	Write the sum over the denominator. Write it in simplest form.

Step 1:

$$\begin{array}{r} \frac{2}{6} \\ +\frac{1}{6} \\ \hline \end{array}$$

Step 2:

$$\begin{array}{rl} \frac{2}{6} & \leftarrow \quad 2 \text{ sixths} \\ +\frac{1}{6} & \leftarrow + \ 1 \text{ sixth} \\ \hline & \quad 3 \text{ sixths} \end{array}$$

Step 3:

$$\begin{array}{r} \frac{2}{6} \\ +\frac{1}{6} \\ \hline \frac{3}{6} = \frac{1}{2} \end{array}$$

So, $\frac{2}{6} + \frac{1}{6} = \frac{1}{2}$.

To subtract like fractions, subtract the numerators.
Remember, the denominator stays the same. Then write the
difference over the denominator.

Find the sum or difference. Write the answer in simplest form.

1. $\frac{1}{5} + \frac{2}{5}$

2. $\frac{3}{7} + \frac{2}{7}$

3. $\frac{4}{9} + \frac{2}{9}$

4. $\frac{8}{9} - \frac{7}{9}$

5. $\frac{7}{8} - \frac{1}{8}$

6. $\frac{9}{12} - \frac{5}{12}$

7. $\frac{2}{6} + \frac{3}{6}$

8. $\frac{1}{8} + \frac{3}{8}$

9. $\frac{6}{10} + \frac{3}{10}$

10. $\frac{6}{8} - \frac{1}{8}$

11. $\frac{4}{6} - \frac{1}{6}$

12. $\frac{7}{14} - \frac{4}{14}$

Name _____

Add Unlike Fractions

Use fraction bars to add fractions.

Show $\frac{1}{3} + \frac{1}{6}$ with fraction bars.

1

$\frac{1}{3}$	$\frac{1}{6}$

Now, find like fraction bars that fit exactly under the sum $\frac{1}{3} + \frac{1}{6}$.

So, three sixth bars fit under the sum.

$\frac{3}{6}$ equals $\frac{1}{2}$ in simplest form.

So, $\frac{1}{3} + \frac{1}{6} = \frac{1}{2}$.

1

$\frac{1}{3}$	$\frac{1}{6}$

$\frac{1}{6}$	$\frac{1}{6}$	$\frac{1}{6}$

Use fraction bars to find the sum.

1.

1

$\frac{1}{2}$	$\frac{1}{8}$	$\frac{1}{8}$	$\frac{1}{8}$

$\frac{7}{8}$

2.

1

$\frac{1}{3}$	$\frac{1}{12}$	$\frac{1}{12}$	$\frac{1}{12}$	$\frac{1}{12}$	$\frac{1}{12}$

$\frac{9}{12}$, or $\frac{3}{4}$

3.

1

$\frac{1}{10}$	$\frac{1}{10}$	$\frac{1}{10}$	$\frac{1}{10}$	$\frac{1}{10}$	$\frac{1}{10}$	$\frac{1}{10}$	$\frac{1}{5}$

$\frac{9}{10}$

4.

1

$\frac{1}{5}$	$\frac{1}{5}$	$\frac{1}{2}$

$\frac{9}{10}$

5.

1

$\frac{1}{6}$	$\frac{1}{3}$	$\frac{1}{3}$

$\frac{5}{6}$

6.

1

$\frac{1}{4}$	$\frac{1}{4}$	$\frac{1}{3}$

$\frac{10}{12}$, or $\frac{5}{6}$

7. $\frac{1}{5} + \frac{1}{2}$

$\frac{7}{10}$

8. $\frac{2}{3} + \frac{3}{4}$

$1\frac{5}{12}$

9. $\frac{2}{6} + \frac{1}{4}$

$\frac{7}{12}$

10. $\frac{1}{3} + \frac{1}{2}$

$\frac{5}{6}$

11. $\frac{3}{5} + \frac{3}{10}$

$\frac{9}{10}$

12. $\frac{1}{4} + \frac{5}{6}$

$1\frac{1}{12}$

13. $\frac{3}{8} + \frac{3}{4}$

$1\frac{1}{8}$

14. $\frac{1}{6} + \frac{2}{3}$

$\frac{5}{6}$

Subtract Unlike Fractions

Use fraction bars to subtract fractions.

Show $\dfrac{5}{6} - \dfrac{1}{3}$ with fraction bars.

Now, find like fraction bars that fit exactly under the difference $\dfrac{5}{6} - \dfrac{1}{3}$.

So, three sixth bars fit under the difference.

$\dfrac{3}{6}$ equals $\dfrac{1}{2}$ in simplest form.

So, $\dfrac{5}{6} - \dfrac{1}{3} = \dfrac{1}{2}$.

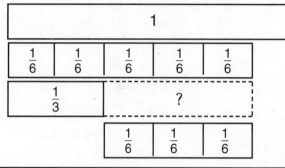

Use fraction bars to find the difference.

1.

$\dfrac{7}{12}$

2.

$\dfrac{2}{10}$, or $\dfrac{1}{5}$

3.

$\dfrac{1}{6}$

4.

$\dfrac{5}{8}$

5.

$\dfrac{1}{6}$

6.

$\dfrac{5}{10}$, or $\dfrac{1}{2}$

7. $\dfrac{3}{4} - \dfrac{1}{2} =$ _____ $\dfrac{1}{4}$

8. $\dfrac{6}{8} - \dfrac{1}{4} =$ _____ $\dfrac{4}{8}$, or $\dfrac{1}{2}$

9. $\dfrac{2}{3} - \dfrac{1}{2} =$ _____ $\dfrac{1}{6}$

Estimate Sums and Differences

You can round fractions to 0, to $\frac{1}{2}$, or to 1 to estimate sums and differences.

Estimate the sum $\frac{3}{5} + \frac{8}{9}$.

Step 1 Find $\frac{3}{5}$ on the number line.
Is it closest to 0, $\frac{1}{2}$, or 1?
The fraction $\frac{3}{5}$ is closest to $\frac{1}{2}$.

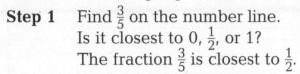

Step 2 Find $\frac{8}{9}$ on the number line.
Is it closest to 0, $\frac{1}{2}$, or 1?
The fraction $\frac{8}{9}$ is closest to 1.

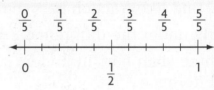

Step 3 To estimate the sum $\frac{3}{5} + \frac{8}{9}$, add the two rounded numbers.

$$\frac{1}{2} + 1 = 1\frac{1}{2}$$

So, $\frac{3}{5} + \frac{8}{9}$ is about $1\frac{1}{2}$.

Use the number lines to estimate whether each fraction is closest to 0, to $\frac{1}{2}$, or to 1. Then find the sum or difference. The first one is done for you.

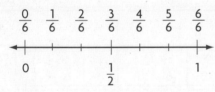
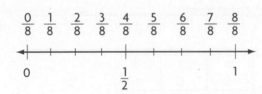

1. $\frac{4}{6} + \frac{1}{8}$

$\boxed{\frac{1}{2}} + \boxed{0}$

$\dfrac{\quad\frac{1}{2}\quad}{}$

2. $\frac{2}{6} + \frac{7}{8}$

$\boxed{\frac{1}{2}} + \boxed{1}$

$\dfrac{\quad 1\frac{1}{2}\quad}{}$

3. $\frac{5}{6} - \frac{3}{8}$

$\boxed{1} - \boxed{\frac{1}{2}}$

$\dfrac{\quad\frac{1}{2}\quad}{}$

4. $\frac{4}{6} + \frac{3}{8}$

$\boxed{\frac{1}{2}} + \boxed{\frac{1}{2}}$

$\dfrac{\quad 1\quad}{}$

5. $\frac{7}{8} - \frac{5}{6}$

$\boxed{1} - \boxed{1}$

$\dfrac{\quad 0\quad}{}$

6. $\frac{1}{6} + \frac{7}{8}$

$\boxed{0} + \boxed{1}$

$\dfrac{\quad 1\quad}{}$

Use Least Common Denominators

Tim and Barbara shared a pizza. Tim ate $\frac{1}{3}$ of the pizza, and Barbara ate $\frac{4}{9}$.

How much of the pizza did they eat in all?

You add $\frac{1}{3} + \frac{4}{9}$ to answer this question.

Use the least common denominator to add the unlike fractions.

The least common denominator is the least common multiple of the denominators.

Step 1: Find the least common multiple of the denominators.

multiples of 3: 3, 6, ⑨, 12

multiples of 9: ⑨, 18, 27

Step 2: Rename each fraction, using the least common denominator.

$\frac{1}{3} = \frac{3}{9}$ and $\frac{4}{9} = \frac{4}{9}$

Step 3: Add the like fractions.

$\frac{3}{9} + \frac{4}{9} = \frac{7}{9}$

Step 4: This sum is already in simplest form.

$\frac{7}{9}$

So, Barbara and Tim ate $\frac{7}{9}$ of the pizza.

To subtract unlike fractions, follow steps 1 and 2. Then subtract and simplify.

Find the least common denominator.

1. $\frac{1}{3} + \frac{1}{5}$

15

2. $\frac{2}{5} + \frac{1}{2}$

10

3. $\frac{3}{6} + \frac{3}{4}$

12

4. $\frac{7}{9} + \frac{5}{6}$

18

Find the sum or difference. Write the answer in simplest form.

5. $\frac{5}{6} + \frac{3}{4} =$ $1\frac{7}{12}$

6. $\frac{7}{12} - \frac{1}{4} =$ $\frac{1}{3}$

7. $\frac{6}{10} + \frac{3}{5} =$ $1\frac{1}{5}$

8. $\frac{5}{6} - \frac{7}{12} =$ $\frac{1}{4}$

9. $\frac{1}{2} + \frac{9}{12} =$ $1\frac{1}{4}$

10. $\frac{3}{4} - \frac{1}{3} =$ $\frac{5}{12}$

© Harcourt

Add and Subtract Unlike Fractions

When you add or subtract two fractions with unlike denominators, you need to make the denominators the same. Find the least common denominator (LCD), and change the fractions to like fractions with that denominator.

Add. $\frac{2}{3} + \frac{1}{4} = n$

Step 1

Find the multiples of both denominators to determine the LCM.

$3 = 3, 6, 9, 12, \ldots$

$4 = 4, 8, 12, 16, \ldots$

The LCM of 3 and 4 is 12. So, the LCD of $\frac{2}{3}$ and $\frac{1}{4}$ is 12.

Step 2

Use the LCD to make like fractions. Multiply the numerator and denominator by the same number.

$\frac{2}{3} = \frac{2 \times 4}{3 \times 4} = \frac{8}{12}$

$+\frac{1}{4} = \frac{1 \times 3}{4 \times 3} = \frac{3}{12}$

Step 3

Add the fractions.

$$\frac{8}{12}$$
$$+\frac{3}{12}$$
$$\overline{\frac{11}{12}}$$

So, $n = \frac{11}{12}$.

So, the sum of $\frac{2}{3} + \frac{1}{4} = \frac{11}{12}$. This answer is in simplest form.

To subtract fractions with unlike denominators, follow these 3 steps. However, in Step 3, subtract the fractions and write the answer in simplest form.

Write like fractions. Then find the sum or difference. Write the answer in simplest form.

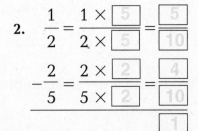

1. $\dfrac{1}{3} = \dfrac{1 \times \boxed{3}}{3 \times \boxed{3}} = \dfrac{\boxed{3}}{\boxed{9}}$

   $+\dfrac{4}{9} \qquad = \dfrac{\boxed{4}}{\boxed{9}}$

   $\qquad\qquad \dfrac{\boxed{7}}{\boxed{9}}$

   Simplest form: ___$\frac{7}{9}$___

2. $\dfrac{1}{2} = \dfrac{1 \times \boxed{5}}{2 \times \boxed{5}} = \dfrac{\boxed{5}}{\boxed{10}}$

   $-\dfrac{2}{5} = \dfrac{2 \times \boxed{2}}{5 \times \boxed{2}} = \dfrac{\boxed{4}}{\boxed{10}}$

   $\qquad\qquad\qquad \dfrac{\boxed{1}}{\boxed{10}}$

   Simplest form: ___$\frac{1}{10}$___

3. $\dfrac{3}{9} = \dfrac{3 \times \boxed{2}}{9 \times \boxed{2}} = \dfrac{\boxed{6}}{\boxed{18}}$

   $+\dfrac{1}{6} = \dfrac{1 \times \boxed{3}}{6 \times \boxed{3}} = \dfrac{\boxed{3}}{\boxed{18}}$

   $\qquad\qquad\qquad \dfrac{\boxed{9}}{\boxed{18}}$

   Simplest form: ___$\frac{1}{2}$___

© Harcourt

Name _____

Problem Solving Strategy

Work Backward

The students in Jason's class started measuring their heights at the beginning of January. By March 1, Jason had grown $\frac{3}{4}$ inch. In February, Jason grew $\frac{3}{8}$ inch. How much did he grow in January?

You can solve the problem by working backward.

Start with the amount he had grown by March 1, and subtract the amount he grew in February.

Find $\frac{3}{4} - \frac{3}{8}$ by using the LCD method.

The LCD of 4 and 8 is 8. Change each fraction into eighths, and subtract the numerators.

$$\frac{3}{4} \times \boxed{\frac{2}{2}} = \frac{6}{8} \qquad \frac{3}{8} \times \boxed{\frac{1}{1}} = \frac{3}{8} \qquad \frac{6}{8} - \frac{3}{8} = \frac{3}{8}$$

So, Jason grew $\frac{3}{8}$ inch in January.

Work backward to solve.

1. Paula is in Jason's class. By March 1, she had grown $\frac{7}{8}$ inch. In February, she grew $\frac{1}{4}$ inch. How much did she grow in January?

   _____ $\frac{5}{8}$ in. _____

2. Sid is in Jason's class. By April 1, he had grown $\frac{15}{16}$ inch. In March, he grew $\frac{1}{8}$ inch, and in February, he grew $\frac{3}{8}$ inch. How much did he grow in January?

   _____ $\frac{7}{16}$ in. _____

3. Harry started the day by trading 5 of his comic books for 7 of Jenny's. Next, he bought 8 at the store. Then he gave Tom 9 comic books. Harry came home with 12 comic books. How many did Harry start the day with?

   _____ 11 comic books _____

4. Wesley started with his favorite number. Then he subtracted 7 from it. He multiplied this difference by 3 and then added 5. Finally, he divided this number by 11. His end result was 1. What is Wesley's favorite number?

   _____ 9 _____

© Harcourt

Add Mixed Numbers

Fred and Gregg are going to put up a tent. They need two pieces of rope to secure the tent. One has to be $3\frac{1}{4}$ feet long and the other $2\frac{1}{2}$ feet long. How much rope do they need?

To find the answer, you must add $3\frac{1}{4} + 2\frac{1}{2}$.

You can add mixed numbers by following these steps.

Step 1

Add the whole numbers. $3 + 2 = 5$

Step 2

Find the LCD. Write equivalent fractions. Add the fractions.

multiples of 4: ④, 8, 12 $\frac{1}{4} + \frac{1}{2} =$

multiples of 2: 2, ④, 6 $\frac{1}{4} + \frac{2}{4} = \frac{3}{4}$

$\frac{1 \times 1}{4 \times 1} = \frac{1}{4} \quad \frac{1 \times 2}{2 \times 2} = \frac{2}{4}$

Step 3

Add the sum of the whole numbers to the sum of the fractions. Write the answer in simplest form if needed.

$5 + \frac{3}{4} = 5\frac{3}{4}$

So, $3\frac{1}{4} + 2\frac{1}{2} = 5\frac{3}{4}$.

Find the sum in simplest form.

1. $3\frac{5}{8}$
    $+2\frac{1}{8}$
    $5\frac{3}{4}$

2. $6\frac{1}{3}$
    $+2\frac{1}{12}$
    $8\frac{5}{12}$

3. $4\frac{1}{4}$
    $+2\frac{1}{4}$
    $6\frac{1}{2}$

4. $5\frac{3}{7}$
    $+1\frac{3}{7}$
    $6\frac{6}{7}$

5. $7\frac{1}{2}$
    $+2\frac{1}{3}$
    $9\frac{5}{6}$

6. $4\frac{3}{5}$
    $+2\frac{1}{10}$
    $6\frac{7}{10}$

7. $4\frac{1}{2}$
    $+3\frac{3}{8}$
    $7\frac{7}{8}$

8. $3\frac{3}{4}$
    $+2\frac{1}{8}$
    $5\frac{7}{8}$

Name _____

Subtract Mixed Numbers

Sonia cut out a pattern for a new skirt from the $3\frac{1}{2}$ yards of material she bought. The pattern used $2\frac{1}{3}$ yards. How much material was left?

You can answer the question by subtracting, $3\frac{1}{2} - 2\frac{1}{3}$.

To subtract mixed numbers, follow these steps.

Step 1

Find the LCD of the fractions by listing the multiples of each number.

Multiples of 2:   2,   4,   ⑥   8,   10

Multiples of 3:   3,   ⑥   9,   12,   15

Since 6 is the first common multiple, it is the least common multiple.

Step 2

Change the fractions into like fractions with 6 as the denominator.

$$\frac{1 \times 3}{2 \times 3} = \frac{3}{6} \qquad \frac{1 \times 2}{3 \times 2} = \frac{2}{6}$$

Step 3

Subtract the fractions.

$$\begin{array}{r} 3\frac{1}{2} = 3\frac{3}{6} \\ -2\frac{1}{3} = -2\frac{2}{6} \\ \hline \frac{1}{6} \end{array}$$

Step 4

Subtract the whole numbers.

$$\begin{array}{r} 3\frac{1}{2} = 3\frac{3}{6} \\ -2\frac{1}{3} = -2\frac{2}{6} \\ \hline 1\frac{1}{6} \end{array}$$

So, Sonia has $1\frac{1}{6}$ yards left.

Subtract. Write the answer in simplest form.

1. $\begin{array}{r} 4\frac{4}{5} = 4\frac{8}{10} \\ -1\frac{1}{10} = -1\frac{1}{10} \\ \hline 3\frac{7}{10} \end{array}$

2. $\begin{array}{r} 6\frac{2}{3} = 6\frac{4}{6} \\ -4\frac{1}{6} = -4\frac{1}{6} \\ \hline 2\frac{1}{2} \end{array}$

3. $\begin{array}{r} 7\frac{3}{4} = 7\frac{9}{12} \\ -4\frac{5}{12} = -4\frac{5}{12} \\ \hline 3\frac{1}{3} \end{array}$

4. $\begin{array}{r} 8\frac{1}{3} = 8\frac{4}{12} \\ -1\frac{1}{4} = -1\frac{3}{12} \\ \hline 7\frac{1}{12} \end{array}$

5. $\begin{array}{r} 2\frac{7}{8} = 2\frac{7}{8} \\ -1\frac{1}{2} = -1\frac{4}{8} \\ \hline 1\frac{3}{8} \end{array}$

6. $\begin{array}{r} 6\frac{7}{9} = 6\frac{7}{9} \\ -4\frac{2}{3} = -4\frac{6}{9} \\ \hline 2\frac{1}{9} \end{array}$

© Harcourt

Subtraction with Renaming

Wayne had $4\frac{1}{4}$ feet of rope. He gave $2\frac{2}{3}$ feet to his friend.
How much rope did he have left?

You can answer the question by subtracting, $4\frac{1}{4} - 2\frac{2}{3}$.

To subtract mixed numbers, follow these steps.

Step 1

Find the LCD.

Multiples of 4: 4, 8, (12,) 16

Multiples of 3: 3, 6, 9, (12)

So, 12 is the LCD.

Step 2

Change each fraction into a fraction with the denominator 12.

$$\frac{1 \times 3}{4 \times 3} = \frac{3}{12}$$

$$\frac{2 \times 4}{3 \times 4} = \frac{8}{12}$$

Step 3

Replace the unlike fractions with the like fractions.

$$4\frac{1}{4} = 4\frac{3}{12}$$
$$-2\frac{2}{3} = -2\frac{8}{12}$$

Step 4

Rename 1 whole from 4 to subtract the fractions. Rename the 1 as $\frac{12}{12}$.

$$4\frac{3}{12} = 3\frac{15}{12} \qquad \frac{12}{12} + \frac{3}{12} = \frac{15}{12}$$
$$-2\frac{8}{12} = -2\frac{8}{12}$$

Step 5

Subtract the fractions.

$$4\frac{3}{12} = 3\frac{15}{12}$$
$$-2\frac{8}{12} = -2\frac{8}{12}$$
$$\frac{7}{12}$$

Step 6

Subtract the whole numbers.

$$4\frac{3}{12} = 3\frac{15}{12}$$
$$-2\frac{8}{12} = -2\frac{8}{12}$$
$$1\frac{7}{12}$$

So, Wayne has $1\frac{7}{12}$ feet left.

Find the difference in simplest form.

1. $5\frac{1}{4}$
    $-\frac{1}{2}$

    $4\frac{3}{4}$

2. $6\frac{1}{8}$
    $-2\frac{1}{4}$

    $3\frac{7}{8}$

3. $5\frac{3}{10}$
    $-1\frac{3}{5}$

    $3\frac{7}{10}$

4. $4\frac{1}{6}$
    $-2\frac{2}{3}$

    $1\frac{1}{2}$

© Harcourt

Name _____

Practice with Mixed Numbers

Larry made $2\frac{5}{6}$ pounds of baked ziti. He and his brother ate $1\frac{1}{3}$ pounds. How much was left over? Use fraction bars to find the answer.

Subtract. $2\frac{5}{6} - 1\frac{1}{3}$ Estimate: about $1\frac{1}{2}$ pound

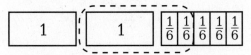

$2\frac{5}{6} - 1\frac{1}{3} = 1\frac{3}{6}$ or $1\frac{1}{2}$ pounds

---

Add or subtract. Write the answer in simplest form. Estimate to check.

1.
$$3\frac{13}{15}$$
$$+2\frac{1}{5}$$
$$6\frac{1}{15}$$

2.
$$1\frac{5}{12}$$
$$+2\frac{1}{6}$$
$$3\frac{7}{12}$$

3.
$$5\frac{3}{4}$$
$$-3\frac{7}{8}$$
$$1\frac{7}{8}$$

4.
$$6\frac{2}{3}$$
$$-1\frac{10}{12}$$
$$4\frac{5}{6}$$

5.
$$2\frac{3}{8}$$
$$+4\frac{7}{8}$$
$$7\frac{1}{4}$$

6.
$$9\frac{4}{5}$$
$$-2\frac{2}{3}$$
$$7\frac{2}{15}$$

7.
$$7\frac{1}{12}$$
$$-2\frac{1}{6}$$
$$4\frac{11}{12}$$

8.
$$4\frac{2}{5}$$
$$+1\frac{1}{3}$$
$$5\frac{11}{15}$$

Algebra Find the value of *n*.

9. $3\frac{1}{2} + n = 5$

$n = 1\frac{1}{2}$
_____

10. $n - 4\frac{1}{8} = 6\frac{1}{2}$

$n = 10\frac{5}{8}$
_____

11. $4\frac{6}{7} - n = 2\frac{1}{7}$

$n = 2\frac{5}{7}$
_____

12. $n + 11\frac{1}{6} = 15\frac{1}{3}$

$n = 4\frac{1}{6}$
_____

Problem Solving Skill: Multistep Problems

Hank bought a piece of wood that was 8 feet long. He used $1\frac{1}{4}$ feet for a shelf in his room and $2\frac{1}{4}$ feet for a shelf in his sister's room. Then he made a box using another $3\frac{1}{4}$ feet. How much of the wood does he have left?

You can solve the problem by doing more than one operation. First add the $1\frac{1}{4}$ feet for his shelf, the $2\frac{1}{4}$ feet for his sister's shelf, and the $3\frac{1}{4}$ feet for his box.

$$
\begin{array}{r}
1\frac{1}{4} \\
2\frac{1}{4} \\
+3\frac{1}{4} \\
\hline
6\frac{3}{4}
\end{array}
$$

Then you would subtract the total amount of $6\frac{3}{4}$ feet from the 8 feet he bought. $8 - 6\frac{3}{4} = 1\frac{1}{4}$

So, Hank has $1\frac{1}{4}$ feet of wood left.

Solve.

1. Ralph bought 12 feet of wood. He made four projects. The first one used $3\frac{1}{2}$ feet, the second one used $2\frac{1}{4}$ feet, the third one used $2\frac{3}{4}$ feet, and the fourth one used $1\frac{1}{4}$ feet. How much wood did he have left?

   $2\frac{1}{4}$ ft

2. Nancy read every day for five days. She read 8 pages on Monday, 12 pages on Tuesday, 25 pages on Thursday, and 40 pages on Friday. If she read a total of 156 pages, how many pages did she read on Wednesday?

   71 pages

3. On Monday Charley drove 32 miles, on Tuesday 58 miles, on Wednesday 88 miles, and on Thursday 94 miles. His total for five days was 335 miles. How far did he drive on Friday?

   63 mi

4. Lacy was serving pizza at a party. She gave the first person $\frac{1}{8}$ of the pizza, the second person $\frac{3}{8}$, and the third person $\frac{1}{4}$ of the pizza. How much of the pizza is left?

   $\frac{1}{4}$ of the pizza

Multiply Fractions and Whole Numbers

Hector had 12 baseball cards. He gave $\frac{2}{3}$ of them to his friend Ned. How many baseball cards did he give to Ned?

You can answer the question by multiplying $\frac{2}{3} \times 12$. To multiply a fraction and a whole number you can use a model:

Step 1 Draw 12 rectangles to show the cards.

Step 2 The denominator of the fraction $\frac{2}{3}$ is 3. This means there are 3 equal parts, so divide the rectangles into 3 equal groups.

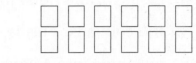
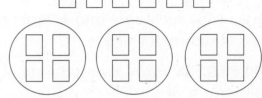

Step 3 The numerator of the fraction $\frac{2}{3}$ is 2. This means there are 2 parts given, so shade 2 of the groups.

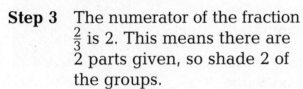

Step 4 Count the shaded rectangles, or cards. There are 8 cards.

So, $\frac{2}{3} \times 12 = 8$.

Write the number sentence each model represents.

1.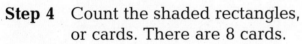

$\frac{2}{5} \times 10 = 4$

2.

$\frac{3}{4} \times 12 = 9$

3.

$\frac{2}{3} \times 9 = 6$

Draw a picture to help you multiply. Find the product. Check students' drawings.

4. $\frac{4}{9} \times 27 = \underline{\quad 12 \quad}$

5. $\frac{1}{6} \times 12 = \underline{\quad 2 \quad}$

6. $\frac{3}{5} \times 20 = \underline{\quad 12 \quad}$

Multiply a Fraction by a Fraction

Multiply. $\frac{3}{4} \times \frac{3}{5}$

To multiply fractions you can use a rectangle model. Follow these guidelines:

- Draw a rectangle, and divide the rectangle into 5 equal columns. This is for the denominator of $\frac{3}{5}$.

- Shade 3 of the columns. This is for the numerator of $\frac{3}{5}$.

- Divide the rectangle into 4 equal rows. This is for the denominator of $\frac{3}{4}$.

- Shade 3 of the rows with diagonal lines. This is for the numerator of $\frac{3}{4}$.

- Count how many pieces the rectangle is divided into. There are 20 pieces. This is the new denominator.

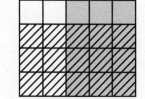

- Count how many pieces have overlapping lines and shading. There are 9. This is the new numerator.

So, $\frac{3}{4} \times \frac{3}{5} = \frac{9}{20}$.

Divide and shade a rectangle model to find the product. Check students' drawings.

1. $\frac{1}{3} \times \frac{5}{6} = \frac{5}{18}$

2. $\frac{5}{8} \times \frac{3}{4} = \frac{15}{32}$

3. $\frac{1}{4} \times \frac{3}{8} = \frac{3}{32}$

4. $\frac{2}{5} \times \frac{1}{3} = \frac{2}{15}$

5. $\frac{1}{2} \times \frac{7}{8} = \frac{7}{16}$

6. $\frac{5}{6} \times \frac{3}{4} = \frac{15}{24}$

7. $\frac{1}{4} \times \frac{5}{6} = \frac{5}{24}$

8. $\frac{2}{3} \times \frac{1}{4} = \frac{2}{12}$

9. $\frac{2}{7} \times \frac{3}{4} = \frac{6}{28}$

10. $\frac{3}{5} \times \frac{3}{5} = \frac{9}{25}$

11. $\frac{4}{5} \times \frac{1}{2} = \frac{4}{10}$

12. $\frac{5}{9} \times \frac{1}{2} = \frac{5}{18}$

Multiply Fractions and Mixed Numbers

Multiply. $\frac{2}{3} \times 2\frac{1}{4}$

You can find the product by using the Distributive Property.
The Distributive Property allows you to break apart numbers
to multiply.

To multiply a fraction and a mixed number, break apart the
mixed number.

$$\frac{2}{3} \times 2\frac{1}{4} = \frac{2}{3} \times \left(2 + \frac{1}{4}\right) \quad \longleftarrow \text{Break apart the mixed number.}$$

$$= \left(\frac{2}{3} \times 2\right) + \left(\frac{2}{3} \times \frac{1}{4}\right) \quad \longleftarrow \text{Multiply each part.}$$

$$= \frac{4}{3} + \frac{2}{12}$$

$$= \frac{16}{12} + \frac{2}{12} \quad \longleftarrow \text{Find the LCD and rename the fractions.}$$

$$= \frac{18}{12} = 1\frac{1}{2} \quad \longleftarrow \text{Add the products. Simplify the sum.}$$

So, $\frac{2}{3} \times 2\frac{1}{4} = 1\frac{1}{2}$.

Multiply. Write the answer in simplest form.

1. $\frac{1}{3} \times 3\frac{1}{5} =$ _____ $\frac{16}{15}$, or $1\frac{1}{15}$ _____

2. $\frac{1}{2} \times 2\frac{3}{4} =$ _____ $\frac{11}{8}$, or $1\frac{3}{8}$ _____

3. $\frac{1}{6} \times 3\frac{2}{3} =$ _____ $\frac{11}{18}$ _____

4. $\frac{1}{4} \times 2\frac{5}{6} =$ _____ $\frac{17}{24}$ _____

5. $\frac{1}{3} \times 3\frac{1}{2} =$ _____ $\frac{7}{6}$, or $1\frac{1}{6}$ _____

6. $\frac{1}{8} \times 4\frac{1}{4} =$ _____ $\frac{17}{32}$ _____

7. $\frac{3}{8} \times 1\frac{1}{4} =$ _____ $\frac{15}{32}$ _____

8. $\frac{4}{5} \times 2\frac{1}{2} =$ _____ $\frac{2}{1}$, or 2 _____

Multiply with Mixed Numbers

Multiply. $1\frac{2}{3} \times 1\frac{1}{2}$

To multiply two mixed numbers, follow the same steps you use to multiply a fraction and a mixed number.

Step 1

Write each mixed number as a fraction.

$$1\frac{2}{3} = \frac{(3 \times 1) + 2}{3} = \frac{5}{3}$$

$$1\frac{1}{2} = \frac{(2 \times 1) + 1}{2} = \frac{3}{2}$$

Step 2

Multiply the fractions,

$$\frac{5 \times 3}{3 \times 2} = \frac{15}{6}$$

or cancel the 3 in the numerator and denominator.

$$\frac{5 \times \overset{1}{\cancel{3}}}{\underset{1}{\cancel{3}} \times 2} = \frac{5}{2}$$

Step 3

Write the product as a mixed number in simplest form.

$$\frac{15}{6} = 2\frac{3}{6} = 2\frac{1}{2}$$

or

$$\frac{5}{2} = 2\frac{1}{2}$$

Multiply. Write the answer in simplest form.

1. $2\frac{1}{2} \times 1\frac{1}{5} =$ _____ $\frac{30}{10}$ or 3

2. $1\frac{1}{3} \times 1\frac{1}{2} =$ _____ $\frac{12}{6}$ or 2

3. $1\frac{1}{2} \times 1\frac{1}{4} =$ _____ $\frac{15}{8}$ or $1\frac{7}{8}$

4. $1\frac{3}{4} \times 3\frac{1}{2} =$ _____ $\frac{49}{8}$ or $6\frac{1}{8}$

5. $6\frac{1}{2} \times 1\frac{3}{5} =$ _____ $\frac{104}{10}$ or $10\frac{2}{5}$

6. $1\frac{2}{3} \times 1\frac{2}{3} =$ _____ $\frac{25}{9}$ or $2\frac{7}{9}$

7. $1\frac{1}{5} \times 1\frac{1}{2} =$ _____ $\frac{18}{10}$ or $1\frac{4}{5}$

8. $2\frac{1}{2} \times 1\frac{3}{5} =$ _____ $\frac{40}{10}$ or 4

Problem Solving Skill

Sequence and Prioritize Information

The Perez family planned an evening event that includes a snack, dinner, dessert, and game time.

There are 6 hours planned for the evening. $\frac{1}{3}$ of the evening's time will be devoted to dinner. $\frac{1}{6}$ of the time will be spent on having a snack. How many hours will be spent on playing games and dessert?

Sequencing the information may help you solve this problem. Start with events for which you have some information.

Event	Time
Dinner	6 hours total $\times \frac{1}{3}$ = 2 hours for dinner
Snack	6 hours total $\times \frac{1}{6}$ = 1 hour for snacks

Now subtract the snack and dinner time to find how much time can be devoted to games and dessert.

Event	Time
Games and Dessert	Total time spent on dinner and snacks = 3 hours 6 hours total − 3 hours for dinner and snacks = 3 hours left for games and dessert

---

Sequence the information by starting with the information you know. Then solve the problem.

1. John drives a total of 350 miles a day. He makes 3 stops. He drives 150 miles to his first stop. From the second stop to the third stop, he drives 75 miles. How many miles does he drive from the first stop to the second stop?

   _____ 125 miles _____

2. Mary spent $45.00 altogether at the store. She bought some food for $32.75 and some school supplies. How much did she spend on school supplies?

   _____ $12.25 _____

Divide Fractions

You can use pictures to model division of fractions.

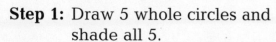

$$5 \div \frac{1}{3} \qquad\qquad\qquad \frac{4}{5} \div \frac{1}{10}$$

Step 1: Draw 5 whole circles and shade all 5.

Step 1: Draw one whole rectangle and shade four fifths of it.

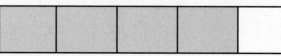

Step 2: Divide each circle into **thirds.**

Step 2: Divide the rectangle into **tenths.**

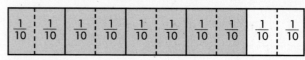

Step 3: Count the number of shaded thirds.

Step 3: Count the number of shaded tenths.

There are 15 thirds in 5. So, $5 \div \frac{1}{3} = 15$.

There are 8 tenths in $\frac{4}{5}$. So, $\frac{4}{5} \div \frac{1}{10} = 8$.

Draw a model for the division problem and find the quotient. Possible models are shown.

1. $\frac{2}{3} \div \frac{1}{9} =$ ___6___

2. $2 \div \frac{1}{5} =$ ___10___

3. $\frac{3}{4} \div \frac{1}{8} =$ ___6___

4. $3 \div \frac{1}{4} =$ ___12___

5. $\frac{1}{2} \div \frac{1}{8} =$ ___4___

6. $\frac{1}{3} \div \frac{1}{6} =$ ___2___

Reciprocals

Reciprocals are two fractions that have a product of 1.

Fractions:

To find the reciprocal of a fraction, switch the numerator and denominator.

The reciprocal of $\frac{3}{8}$ is $\frac{8}{3}$.

$$\frac{3}{8} \times \frac{8}{3} = \frac{24}{24} = 1$$

Whole Numbers:

To find the reciprocal of a whole number, first write it as a fraction. Then switch the numerator and denominator.

$$7 = \frac{7}{1}.$$

The reciprocal of $\frac{7}{1}$ is $\frac{1}{7}$.

Mixed Numbers:

To find the reciprocal of a mixed number, first write it as a fraction. Then switch the numerator and denominator.

$$5\frac{2}{3} = \frac{17}{3}$$

The reciprocal of $\frac{17}{3}$ is $\frac{3}{17}$.

Are the two numbers reciprocals? Write *yes* or *no*.

1. $\frac{1}{9}$ and 19

no

2. $\frac{3}{10}$ and $\frac{10}{3}$

yes

3. $1\frac{3}{5}$ and $\frac{8}{5}$

no

4. 5 and $\frac{1}{5}$

yes

5. $\frac{5}{13}$ and $2\frac{3}{5}$

yes

6. $\frac{1}{10}$ and $\frac{1}{10}$

no

7. $2\frac{1}{4}$ and $\frac{4}{9}$

yes

8. $\frac{7}{12}$ and $\frac{12}{7}$

yes

Write the reciprocal of each number.

9. $\frac{1}{7}$

$\frac{7}{1} = 7$

10. $\frac{5}{12}$

$\frac{12}{5} = 2\frac{2}{5}$

11. 6

$\frac{1}{6}$

12. $3\frac{5}{9}$

$\frac{9}{32}$

13. $\frac{6}{5}$

$\frac{5}{6}$

14. $\frac{2}{11}$

$\frac{11}{2} = 5\frac{1}{2}$

15. 11

$\frac{1}{11}$

16. $1\frac{3}{8}$

$\frac{8}{11}$

17. $\frac{1}{2}$

2

18. 100

$\frac{1}{100}$

Divide Whole Numbers by Fractions

Beth is working on a science project. She needs pieces of wire that are $\frac{2}{3}$ yd long for the project. She bought a piece of wire that is 6 yd long at the hardware store.

How many $\frac{2}{3}$ pieces can she cut from this 6-yd piece?

Step 1: Write a division sentence to find this amount.

$\frac{6}{1} \div \frac{2}{3}$

Think: Write 6 as $\frac{6}{1}$.

Step 2: Use the reciprocal of the divisor to write a multiplication problem.

$\frac{6}{1} \times \frac{3}{2}$

Think: The reciprocal of $\frac{2}{3}$ is $\frac{3}{2}$.

Step 3: Multiply.

$\frac{6}{1} \times \frac{3}{2} = \frac{18}{2} = 9$

So, Beth can cut 9 pieces of wire.

Use the reciprocal to write a multiplication problem. Solve the problem. Write the answer in simplest form. Check multiplication problems.

1. $3 \div \frac{1}{8}$

$\frac{3}{1} \times \frac{8}{1} = 24$

2. $5 \div \frac{1}{2}$

10

3. $10 \div \frac{2}{3}$

15

4. $27 \div \frac{3}{5}$

45

5. $12 \div \frac{4}{5}$

15

6. $8 \div \frac{3}{4}$

$\frac{32}{3} = 10\frac{2}{3}$

7. $18 \div \frac{3}{8}$

48

8. $7 \div \frac{4}{5}$

$\frac{35}{4} = 8\frac{3}{4}$

9. $6 \div \frac{3}{4}$

8

10. $16 \div \frac{4}{5}$

20

11. $9 \div \frac{6}{7}$

$\frac{63}{6} = 10\frac{1}{2}$

12. $2 \div \frac{3}{10}$

$\frac{20}{3} = 6\frac{2}{3}$

13. $9 \div \frac{3}{8}$

24

14. $9 \div \frac{1}{5}$

45

15. $6 \div \frac{3}{20}$

40

16. $20 \div \frac{4}{5}$

25

© Harcourt

Divide Fractions

Connie is working on a craft project. She needs $\frac{3}{8}$-yd pieces of ribbon for the project. She bought a $\frac{3}{4}$-yd piece of ribbon at the craft store.

How many $\frac{3}{8}$-yd pieces can she cut from $\frac{3}{4}$-yd piece?

Step 1: Write a division sentence to find this amount.

$$\frac{3}{4} \div \frac{3}{8}$$

Step 2: Use the reciprocal of the divisor to write a multiplication problem.

$$\frac{3}{4} \times \frac{8}{3}$$

Think: The reciprocal of $\frac{3}{8}$ is $\frac{8}{3}$.

Step 3: Multiply.

$$\frac{3}{4} \times \frac{8}{3} = \frac{24}{12} = 2$$

So, Connie can cut 2 pieces of ribbon.

Use the reciprocal to write a multiplication problem. Solve the problem. Write the answer in simplest form.

1. $\frac{3}{8} \div 24$

$\frac{1}{64}$

2. $\frac{5}{9} \div \frac{2}{3}$

$\frac{5}{6}$

3. $\frac{4}{5} \div \frac{2}{3}$

$\frac{6}{5} = 1\frac{1}{5}$

4. $\frac{5}{12} \div \frac{5}{8}$

$\frac{2}{3}$

5. $\frac{5}{6} \div \frac{1}{3}$

$\frac{15}{6} = 2\frac{1}{2}$

6. $\frac{5}{8} \div \frac{3}{4}$

$\frac{5}{6}$

7. $\frac{4}{5} \div 6$

$\frac{2}{15}$

8. $1\frac{1}{15} \div 7$

$\frac{16}{105}$

9. $2\frac{1}{4} \div \frac{1}{3}$

$6\frac{3}{4}$

10. $1\frac{1}{4} \div 2\frac{1}{3}$

$\frac{15}{28}$

11. $\frac{1}{3} \div \frac{1}{2}$

$\frac{2}{3}$

12. $1\frac{1}{3} \div 1\frac{1}{2}$

$\frac{8}{9}$

13. $\frac{1}{2} \div \frac{1}{4}$

2

14. $\frac{3}{4} \div 1\frac{1}{4}$

$\frac{3}{5}$

15. $\frac{5}{6} \div \frac{1}{3}$

$2\frac{1}{2}$

16. $1\frac{2}{3} \div \frac{1}{3}$

5

Problem Solving Strategy

Solve a Simpler Problem

The bank gave Jim a loan of $4,000. This is $\frac{1}{8}$ of the amount they gave him last year. How much did the bank loan Jim last year?

You can solve a more difficult problem by first solving a simpler one.

Step 1: If you can, change the numbers so that they are easier to work with.

Let 4 represent 4,000.

Step 2: Write the problem, using the new number.

$4 \div \frac{1}{8} = \frac{4}{1} \times \frac{8}{1}$ **Think:** The reciprocal of $\frac{1}{8}$ is $\frac{8}{1}$.

Step 3: Solve the problem, using the new number.

$\frac{4}{1} \times \frac{8}{1} = 32$

Step 4: Adjust the answer, using the original number.

Multiply the answer by 1,000 to adjust.

So, $32 \times 1,000 = 32,000$.

So, the bank loaned Jim $32,000 last year.

Use a simpler problem to solve. Then adjust your answer.

1. Charles spent $600 on a new bike. This was $\frac{2}{3}$ of his savings. How much money was in his savings?

_____ $900

2. The distance from Barbara's house to Raymond's house is 3,200 miles. You can travel $\frac{3}{4}$ of the distance by highway. How many miles cannot be traveled by highway?

_____ 800 mi

Integers

Integers are whole numbers and their opposites.

The positive integers are to the right of 0. The negative integers are to the left of 0. 0 is neither positive nor negative.

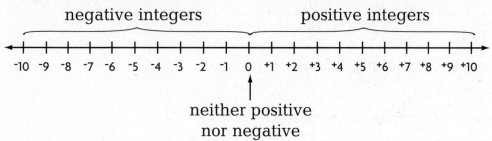

neither positive
nor negative

How far from 0 is $^+5$ and in what direction? _____ 5 units to the right

How far from 0 is $^-5$ and in what direction? _____ 5 units to the left

$^+5$ and $^-5$ are **opposites.** They are the same distance from 0 on a number line, but in opposite directions. Some other opposites are $^+1$ and $^-1$, $^+6$ and $^-6$.

The distance a number is from 0 is referred to as its **absolute value.** $^+5$ and $^-5$ are both 5 units from 0. So, $\left|^-5\right|$ and $\left|^+5\right|$ both equal 5.

The symbol || means absolute value.

How far from 0 is $^-8$? 8 units
So, $\left|^-8\right| = 8$ because it is 8 units from 0.

Name the integer which describes each situation below.

1. an increase in price of $30.00

_____ $^+30$

2. 10 minutes before school starts

_____ $^-10$

3. 4 feet above ground

_____ $^+4$

Write the opposite of each integer.

4. $^+6$ _____ $^-6$

5. $^-28$ _____ $^+28$

6. $^+1,489$ _____ $^-1,489$

7. $^-2,000$ _____ $^+2,000$

Name each integer's absolute value.

8. $\left|^+34\right|$ _____ 34

9. $\left|^-30\right|$ _____ 30

10. $\left|^-235\right|$ _____ 235

Compare and Order Integers

Integers increase as you move right on a number line and decrease as you move left.

Compare ⁻7 to ⁻8. Use <, > or =.

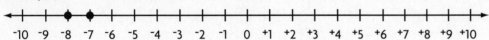

The numbers increase as you move right, and ⁻7 is to the right of ⁻8. So, ⁻7 > ⁻8.

Order ⁺5, ⁻5, ⁻3, and ⁺7 from least to greatest.

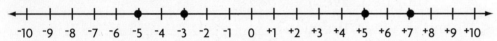

Look at the number line. Since the numbers increase as you move right on the number line, the order from least to greatest is ⁻5, ⁻3, ⁺5, ⁺7.

Name the integer that is 1 *less* than ⁻8.

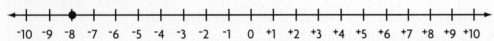

1 less means a decrease, and decreasing amounts move left on a number line. So, 1 to the left of ⁻8 is ⁻9.

Compare. Write <, >, or = for each ◯.

1. ⁺6 ⬤= ⁺6   2. ⁻7 ⬤< ⁻2   3. ⁺3 ⬤> ⁻1   4. ⁻10 ⬤< ⁺8

Order each set of integers from greatest to least.

5. ⁻9, ⁺1, 0, ⁻4   6. ⁺2, ⁻5, ⁺6, ⁻4   7. ⁺10, ⁻10, ⁺1, ⁻1   8. ⁻4, ⁺2, ⁻5, ⁻1

  ⁺1, 0, ⁻4, ⁻9   ⁺6, ⁺2, ⁻4, ⁻5   ⁺10, ⁺1, ⁻1, ⁻10   ⁺2, ⁻1, ⁻4, ⁻5

Name the integer that is one *less* than the given integer.

9. ⁺6 ⁺5   10. ⁻4 ⁻5   11. 0 ⁻1   12. ⁻21 ⁻22   13. ⁺25 ⁺24

Name the integer that is one *more* than the given integer.

14. ⁻10 ⁻9   15. ⁺4 ⁺5   16. ⁻3 ⁻2   17. ⁺32 ⁺33   18. ⁻1 0

Add Integers

This pan balance "weighs" positive and negative numbers. Negative numbers go on the left of the balance and positive numbers go on the right.

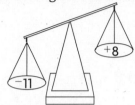

Find ⁻11 + ⁺8.

The scale will tip to the left side, because it is ⁻3 "heavier."

Find ⁻2 + ⁺7.

The scale will tip to the right side, because it is ⁺5 "heavier."

Find ⁻1 + ⁻3.

Both ⁻1 and ⁻3 go on the left side. The scale will tip to the left side, because it is ⁻4 "heavier."

Find how much "heavier" the lower side is.

1.

_____ +5

2.

_____ ⁻4

3.

_____ ⁺1

4.

_____ 0

5.

_____ ⁺5

6.

_____ ⁻14

Solve.

7. ⁺7 + ⁻4

_____ ⁺3

8. ⁺10 + ⁻4

_____ ⁺6

9. ⁻6 + 0

_____ ⁻6

10. ⁻5 + ⁻4

_____ ⁻9

11. ⁻9 + ⁺3

_____ ⁻6

12. ⁺9 + ⁻1

_____ ⁺8

13. ⁺5 + ⁻3 + ⁻2

_____ 0

14. ⁻3 + ⁺5

_____ ⁺2

Reasoning Without adding, tell whether the sum will be negative, positive, or zero.

15. ⁻18 + ⁺25

_____ positive

16. ⁺9 + ⁻20

_____ negative

17. ⁺427 + ⁻427

_____ zero

18. ⁺75 + ⁻19

_____ positive

Subtract Integers

You can use drawings to subtract integers.

Use circles with + signs to represent positive integers and circles with − signs to represent negative integers.

Find $^{+}8 - {}^{+}6$.

Step 1

First, make a drawing of $^{+}8$.

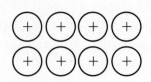

Step 2

To subtract $^{+}6$, take away 6 of the + circles.

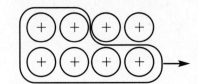

Step 3

The number of circles left represents the difference.

So, $^{+}8 - {}^{+}6 = {}^{+}2$.

Find $^{-}5 - {}^{+}3$.

Step 1

First, make a drawing of $^{-}5$.

Step 2

You cannot subtract $^{+}3$ until you add positive circles.

Add both the positive number and its opposite; $^{+}3$ and $^{-}3$.

Step 3

To subtract $^{+}3$, take away 3 of the + circles.

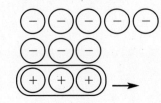

Step 4

The number of circles left represents the difference.

So, $^{-}5 - {}^{+}3 = {}^{-}8$.

Solve.

1. $^{+}5 - {}^{+}3$

   $^{+}2$

2. $^{-}6 - {}^{+}2$

   $^{-}8$

3. $^{+}7 - {}^{+}3$

   $^{+}4$

4. $^{-}9 - {}^{+}2$

   $^{-}11$

5. $^{-}8 - {}^{+}1$

   $^{-}9$

6. $^{+}10 - {}^{+}4$

   $^{+}6$

7. $^{-}4 - {}^{+}4$

   $^{-}8$

8. $^{-}7 - {}^{+}6$

   $^{-}13$

9. $^{-}5 - {}^{+}4$

   $^{-}9$

10. $^{+}8 - {}^{+}2$

    $^{+}6$

11. $^{-}7 - {}^{+}6$

    $^{-}13$

12. $^{-}9 - {}^{+}3$

    $^{-}12$

Subtract Integers

When you subtract a positive integer from a negative integer, you add the opposite. Replace the negative integer with its opposite, and change the operation to addition.

Example: Find $^-6 - {}^+2$.

Step 1 Find the opposite of the number being subtracted. The opposite of $^+2$ is $^-2$.

Step 2 Add the opposite. Change the operation to addition and replace the number being subtracted with its opposite.
$^-6 - {}^+2 = {}^-6 + {}^-2$

Step 3 Add the negative numbers.

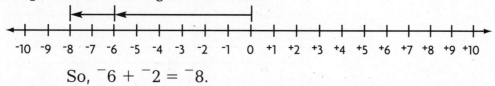

$$^-10 \quad ^-9 \quad ^-8 \quad ^-7 \quad ^-6 \quad ^-5 \quad ^-4 \quad ^-3 \quad ^-2 \quad ^-1 \quad 0 \quad ^+1 \quad ^+2 \quad ^+3 \quad ^+4 \quad ^+5 \quad ^+6 \quad ^+7 \quad ^+8 \quad ^+9 \quad ^+10$$

So, $^-6 + {}^-2 = {}^-8$.

Solve.

1. $^-5 - {}^+3$
_____ $^-8$

2. $^-4 - {}^+7$
_____ $^-11$

3. $^-2 - {}^+8$
_____ $^-10$

4. $^-7 - {}^+1$
_____ $^-8$

5. $^-2 - {}^+1$
_____ $^-3$

6. $^-4 - {}^+6$
_____ $^-10$

7. $^-7 - {}^+3$
_____ $^-10$

8. $0 - {}^+8$
_____ $^-8$

9. $^-9 - {}^+2$
_____ $^-11$

10. $^-5 - {}^+5$
_____ $^-10$

11. $^-2 - {}^+4$
_____ $^-6$

12. $^-5 - {}^+1$
_____ $^-6$

Algebra Complete.

13. $^-2 - {}^+6 = {}^-2 + \boxed{^-6}$

14. $^-7 - {}^+7 = {}^-7 + \boxed{^-7}$

15. $^-3 - {}^+9 = {}^-3 + \boxed{^-9}$

16. $^-5 - {}^+5 = {}^-5 + \boxed{^-5}$

Compare. Write $<$, $>$, or $=$ in each $\bigcirc$.

17. $^-5 - {}^+6 \;\boxed{<}\; ^-2 - {}^+6$

18. $2 - {}^+8 \;\boxed{<}\; 5 + {}^-7$

19. $^-3 - {}^+7 \;\boxed{=}\; ^-3 + {}^-7$

20. $^-5 - {}^+2 \;\boxed{<}\; ^-6 + {}^+1$

Problem Solving Strategy

Draw a Diagram

Draw a diagram to solve.

PROBLEM: Erik and his friends are practicing scuba diving in a 20-foot-deep, 30-foot-long pool for class. First, they had to go down 15 feet. Then they had to go down 2 more feet to practice clearing the water out of their masks. Then they went up 9 feet and back down 5 feet. At what depth are they now?

When solving problems involving integers, first look for key words to determine the positive and negative numbers. A few key words are listed.

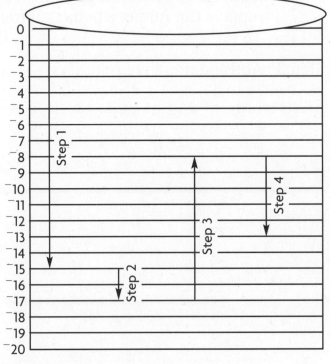

negative (−)	positive (+)
down	up
below	above
drop	raise

The key words in the problem are *up* (+) and *down* (−). Use the key words to step out the problem.

Step 1 down 15 feet (⁻15)

Step 2 down 2 more feet (⁻2)

Step 3 up 9 feet (⁺9)

Step 4 down 5 feet (⁻5)

Making or using a diagram will help you visualize the steps to solve the problem. Look at the diagram.

The last step shows that Erik and his friends will go from ⁻8 feet down 5 more (⁻5) feet. The students are now at 13 feet below the surface, or ⁻13 feet.

Draw a diagram to solve. Check students' diagrams.

1. The next day a new set of divers were practicing in the pool. They began by diving 19 feet. Then they rose 9 feet, went back down 7 feet and up 5 feet. Where are they now?

   ___12 feet below the surface,___

   ___or ⁻12 feet___

2. Jan went swimming. She dove 15 feet, came up 8 feet, went down 1 foot, and came back up 5 more feet. Where is she now?

   ___3 feet below the surface,___

   ___or ⁻3 feet___

© Harcourt

Name _____

Graph Relationships

Number of dollars, x	1	2	3	4
Number of rolls of pennies, y	2	4	6	8

Bill put his collection of pennies in $0.50 rolls. Every two rolls held $1. He made a table to show the relationship between number of dollars and number of rolls of pennies.

Bill wrote the data as ordered pairs: (1,2), (2,4), (3,6), and (4,8). Then he graphed the points and drew a line to connect them.

The ordered pair (2,4) means that Bill has $2 if he has 4 rolls of pennies.

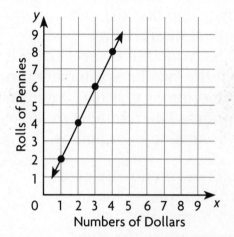

Write the ordered pairs. Then graph the ordered pairs. Check students' work.

1.

Input, x	1	2	3	4
Output, y	4	8	12	16

(1,4), (2,8), (3,12), (4,16)

2.

Input, x	8	10	12	14
Output, y	4	5	6	7

(8,4), (10,5), (12,6), (14,7)

3.

Input, x	2	4	6	8
Output, y	3	5	7	9

(2,3), (4,5), (6,7), (8,9)

4.

Input, x	8	7	6	5
Output, y	6	5	4	3

(8,6), (7,5), (6,4), (5,3)

5. In the problem with Bill's pennies, what does the ordered pair (3,6) mean?

He has $3 with 6 rolls of pennies.

6. In the problem with Bill's pennies, what would be the next ordered pair?

(5,10)

7. How did you decide the answer for problem 6? I looked at the pattern of ordered numbers. The next number for x would be 5 and for y would be 10.

Graph Integers on the Coordinate Plane

A coordinate plane is formed by a horizontal number line (x-axis) and a vertical number line (y-axis), which intersect. The point at which the two lines intersect is named by the ordered pair (0,0) and is called the *origin*. The numbers in the ordered pair are called *coordinates*.

For (⁻2,⁺4), move 2 units left on the x-axis and 4 units up on the y-axis.

For (⁺2,⁺4), move 2 units right on the x-axis and 4 units up on the y-axis.

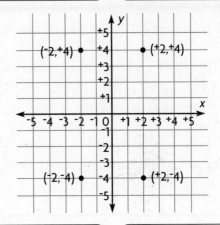

To plot ordered pairs on a coordinate plane, begin at the origin. Positive numbers are to the right and above (0,0). Negative numbers are to the left and below (0,0).

For (⁻2,⁻4), move 2 units left on the x-axis and 4 units down on the y-axis.

For (⁺2,⁻4), move 2 units right on the x-axis and 4 units down on the y-axis.

Write the ordered pair described. Then plot and label the point on the coordinate plane.

1. Start at the origin. Move right 5 units and up 3 units. __(⁺5, ⁺3)__

2. Start at the origin. Move left 4 units and up 1 unit. __(⁻4, ⁺1)__

3. Start at the origin. Move right 2 units and down 3 units. __(⁺2, ⁻3)__

4. Start at the origin. Move left 1 unit and down 3 units. __(⁻1, ⁻3)__

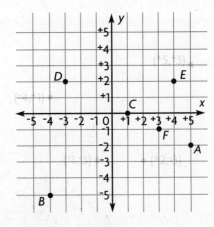

Identify the ordered pair for each point on the coordinate plane above.

5. Point *A*

__(⁺5, ⁻2)__

6. Point *B*

__(⁻4, ⁻5)__

7. Point *C*

__(⁺1,0)__

8. Point *D*

__(⁻3, ⁺2)__

9. Point *E*

__(⁺4, ⁺2)__

10. Point *F*

__(⁺3, ⁻1)__

© Harcourt

Use an Equation to Graph

Sally earns $3 for each hour that she baby-sits. She can graph how much she earns according to the number of hours she baby-sits.

Sally can show this relationship with x and y values in a function table.

Hours, x	1	2	3	4	5
Earns, y	$3	$6	$9	$12	$15

This table shows the amount of money that Sally can earn for the hours she baby-sits. There is a pattern. Find the rule and write the equation that shows the relationship between x and y.

Rule: Multiply x by 3.   Equation: $y = 3x$

Sally knows that for any number of hours, she earns three times that number.

She can find out exactly how much she will make by substituting the number of hours (x) and multiplying it by 3 to get y, how much she earns.

Use a rule to complete the table. Then write the equation.

1.

Table, x	1	2	3	4
Table legs, y	4	8	12	16

$y = 4x$

2.

Fingers, x	5	10	15	20
Hands, y	1	2	3	4

$y = x \div 5$

3.

Input, x	2	3	4	7
Output, y	4	5	6	9

$y = x + 2$

4.

child, x	1	2	3	4
eyes, y	2	4	6	8

$y = 2x$

Use each equation to make a table, write 4 ordered pairs, and then make a graph.
Check students' work. Possible answers are given.

5. $y = x + 4$

x	1	2	3	4
y	5	6	7	8

6. $y = x \div 2$

x	2	4	6	8
y	1	2	3	4

7. $y = x + {}^-4$

x	4	5	6	7
y	0	1	2	3

Problem Solving Skill

Relevant or Irrelevant Information

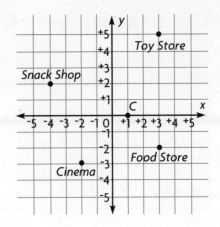

Jonathan gave a map to his visiting cousin so she would be able to find the places she needs. She is looking for the Pet Store and knows that its x-coordinate is the same as the Food Store's x-coordinate. The Snack Shop is north of the Cinema. Jonathan said that the Pet Store is 4 blocks south of the Toy Store. Can you help her find the Pet Store?

Step 1

Decide what you are trying to find.   the coordinates of the Pet Store

Step 2

Read each fact and decide whether it is relevant or irrelevant to solving the problem.

- The Pet Store has the   *relevant*
  the same x-coordinate
  as the Food Store.

- The Snack Shop is   *irrelevant*
  north of the Cinema.

- The Pet Store is 4 blocks   *relevant*
  south of the Toy Store.

Step 3

Use the relevant information to solve the problem.

- The Food Store's x-coordinate is $^+3$.
- The Toy Store's y-coordinate is $^+5$.
- So the Pet Store is at $(^+3, ^+1)$.

---

1. A group of 72 students visited the science center. One third of them visited the planetarium. One half of that number went to the weather exhibit. The remaining students visited the electricity exhibit. Most of the students liked the science center. How many students saw the electricity exhibit?

   36 students

2. Which information is relevant to this problem?

   One third went to the planetarium;

   one half of one third went to the weather exhibit;

   remainder went to the electricity exhibit.

Lines and Angles

In geometry, objects have special names.

You can make lines by connecting any two points. Lines go on forever. You show this by putting arrows at the ends of the line.

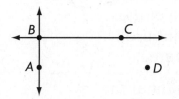

line AB, or $\overleftrightarrow{AB}$, and line BC, or $\overleftrightarrow{BC}$

A **line** is a straight path in a plane. It has no ends. It can be named by any two points on the line.

You can make line segments by joining two points. Line segments do not go on forever. They do not have arrows at the ends.

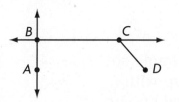

line segment CD, or $\overline{CD}$

A **line segment** is part of a line. It is the shortest distance between two points on a line.

$\overleftrightarrow{CD}$ and $\overleftrightarrow{AD}$ cross each other at point D.

$\overleftrightarrow{AB}$ and $\overleftrightarrow{AD}$ intersect to form right angles.

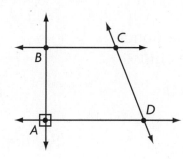

$\overleftrightarrow{BC}$ and $\overleftrightarrow{AD}$ go on forever, and they will never cross.

Lines that cross at one point are **intersecting**.

Lines that intersect to form four right angles are **perpendicular**.

Lines in a plane that never intersect and are the same distance from each other are **parallel**.

Draw and label each object. Check students' drawings.

1. lines AB and CD parallel to each other

2. line segment KL

3. line FG

4. lines EF and GH intersecting at point A

5. lines NO and QR perpendicular to each other

6. lines HI and JK parallel to each other

Measure and Draw Angles

You can use a protractor to measure the angle at the right. A protractor is a tool for measuring the size of the opening of an angle. The unit used to measure an angle is a degree.

A protractor has a center point at the bottom where two lines form right angles. To the right of this is the 0° mark. To the left is the 180° mark.

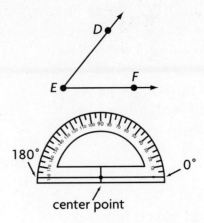

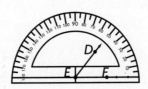

center point

Step 1

Place the protractor on the angle so that the center point lines up with the vertex and the horizontal line on the protractor lines up with ray *EF*.

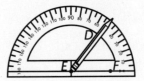

Step 2

To measure the angle, place a pencil on top of the other ray of the angle.

Read the number of degrees the pencil is pointing to.

So, the measure of ∠*DEF* is 50°.

Use a protractor to measure and classify the angle.

1.

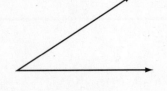

32°; Acute

2.

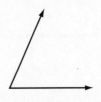

65°; Acute

3.

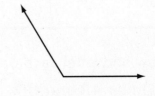

120°; Obtuse

4.

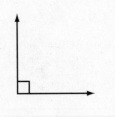

90°; Right

5.

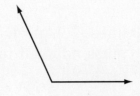

115°; Obtuse

6.

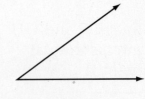

35°; Acute

© Harcourt

Angles and Polygons

How can you remember polygons and their angles? One way is to learn the meanings of the words that describe each shape. Remembering other words that use the same roots can also help you remember the figures.

triangle   *tri-* means 3
   tricycle— a 3-wheeled bicycle

quadrilateral   *quad-* means 4
   quadruplets — 4 babies born at once to the same mother

pentagon   *pent-* means 5
   the Pentagon — a 5-sided building in Washington, D.C.

hexagon   *hex-* means 6
   hex sign — Pennsylvania Dutch art that uses 6-sided figures drawn inside a circle

octagon   *oct-* means 8
   octopus — an animal with 8 legs
   October — used to be the 8th month on the calendar

polygon   *poly-* means many
   polyhedra — a term used for the many-sided shapes of crystals

---

Name each polygon.

1.
_____octagon_____

2.
_____triangle_____

3.
_____hexagon_____

4.
_____octagon_____

5.
_____triangle_____

6.
_____pentagon_____

7.
_____quadrilateral_____

8.
_____hexagon_____

Many quadrilateral shapes have their own meanings as well. Look in the dictionary and find these meanings. Find another word that can help you remember the meaning. Then draw the shape. Check students' drawings.

9. trapezoid   10. parallelogram   11. rectangle   12. rhombus

Circles

You need a centimeter ruler and a compass to construct
a circle with a radius of 2 cm.

A **radius** is a line segment that connects the center with a point on the circle.

A **diameter** is a chord that passes through the center of a circle.

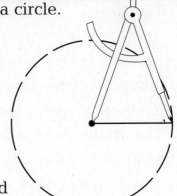

• Draw a point at the center of the circle.

• Start at the point. Use a centimeter ruler to draw a
  line segment 2 cm long. This is the radius.

• Place the point of the compass on the center point.
  Place the pencil point on the other end of the radius.

• Hold the compass point still. Turn the compass around
  on the point to make a complete circle.

Use a centimeter ruler and a compass to construct each circle. Check students'
circles.

1. radius = 1 cm

2. radius = 1.5 cm

3. radius = 2 cm

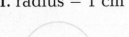

4. diameter = 1.0 cm

5. diameter = 2.4 cm

6. diameter = 5.0 cm

© Harcourt

Name _____

Congruent and Similar Figures

Two figures are similar if they have the same shape.

Two figures are congruent if their matching sides and angles are the same.

To determine if triangles *ABC* and *DEF* are congruent:

- Measure the sizes of the matching angles to see if they are equal.
- Measure the lengths of the matching sides to see if they are equal.

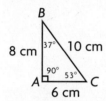

Lengths of Sides
AB = *DE*, *BC* = *EF*, and *AC* = *DF*

Angles
∠*A* = ∠*D*, ∠*B* = ∠*E*, and ∠*C* = ∠*F*

The matching sides are equal, and the matching angles are equal. So, the two triangles are congruent.

Find one pair of similar figures and four pairs of congruent figures.

A. 9 in. 9 in.
B. 9 in. 9 in.
C. 9 in. 8 in.
D. 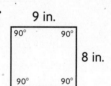 8 in. 9 in.

E. 9 in. 9 in.
F. 8 in. 9 in.
G. 8 in. 9 in.

H. 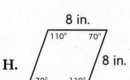 8 in. 8 in.
I. 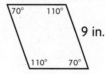 9 in. 9 in.
J. 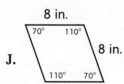 8 in. 8 in.

E and J are similar; A and B, D and F, C and G,

and H and J are congruent

Symmetric Figures

A figure has line symmetry when it can be folded on a line
so that its two parts match. The two halves of the pentagon
are congruent.

Trace the pentagon. Fold it in half along the dotted line. The left
half is congruent to the right half. A figure can have more than one
line of symmetry. Find all the lines of symmetry for the pentagon.

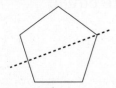   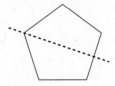

The pentagon has five lines of symmetry in all.

Draw the lines of symmetry. How many lines of symmetry does
each figure have? Check students' drawings.

1.

1 line of symmetry

2.

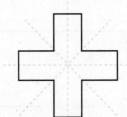

4 lines of symmetry

3.

1 line of symmetry

4.

4 lines of symmetry

5.

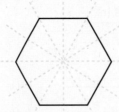

6 lines of symmetry

6.

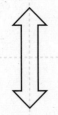

2 lines of symmetry

© Harcourt

Problem Solving Strategy

Find a Pattern

Leonardo Fibonacci was one of the most talented mathematicians of the Middle Ages. One of his hobbies was studying number patterns. One of his most famous patterns is shown below. What is the next term in the pattern?

1, 1, 2, 3, 5, 8

Step 1 What does the problem ask? It asks what the next term in the number pattern is.

Step 2 Find a pattern. The next term is the sum of the two previous terms.

$1 + 1 = 2$, $1 + 2 = 3$, $2 + 3 = 5$, $3 + 5 = 8$

Step 3 Use this information to solve the problem.

$5 + 8 = 13$.   13 is the next term in the pattern.

Find a pattern to solve.

1. What is the next shape in this pattern?

   ○△○△△△○△△△

   _____ circle, _____

2. When Fred's number is 1, Ann's number is 3. When Fred's number is 2, Ann's number is 5. If Fred's is 6, what is Ann's number?

   _____ Ann's number is _____

   _____ 13 _____

3. Write a rule for the pattern described in Problem 2.

   _____ Possible answer: Take _____

   _____ Fred's number, multiply _____

   _____ it by 2 and add 1. _____

4. Alex read 45 pages on Sunday, 90 pages on Monday and 135 pages on Tuesday. If he continues this pattern, how many pages will he read on Friday?

   _____ 270 pages _____

© Harcourt

Triangles

Triangles are polygons with 3 sides and 3 angles. One method of classifying triangles is by the lengths of their sides.

To classify a triangle using this method, you need to know the lengths of its sides.

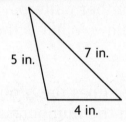

3 congruent sides = **equilateral** triangle

2 congruent sides = **isosceles** triangle

0 congruent sides = **scalene** triangle

Each side is a different length, so this is a scalene triangle.

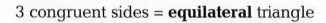

List the number of congruent sides. Then name each triangle. Write *isosceles*, *scalene*, or *equilateral*.

1.
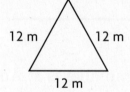
12 m   12 m
12 m

3 sides; equilateral

2.
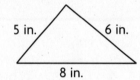
5 in.   6 in.
8 in.

0 sides; scalene

3.
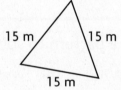
15 m   15 m
15 m

3 sides; equilateral

4.
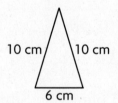
10 cm   10 cm
6 cm

2 sides; isosceles

5.
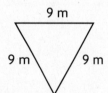
9 m
9 m   9 m

3 sides; equilateral

6.
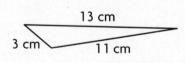
13 cm
3 cm   11 cm

0 sides; scalene

7.
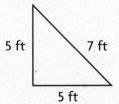
5 ft   7 ft
5 ft

2 sides; isosceles

8.
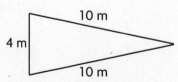
10 m
4 m
10 m

2 sides; isosceles

9.
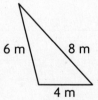
6 m   8 m
4 m

0 sides; scalene

© Harcourt

Name _____

Quadrilaterals

Polygons that have 4 sides and 4 angles are **quadrilaterals**. Quadrilaterals can be classified by looking at the number of parallel sides, the lengths of their sides, and the measures of their angles.

	Trapezoid	Parallelogram	Rectangle	Rhombus	Square
Number of parallel sides	1 pair	2 pairs	2 pairs	2 pairs	2 pairs
Number of congruent sides	0 pair	2 pairs	2 pairs	all 4 sides	all 4 sides
Number of congruent angles	0 pair	2 pairs	all 4 angles	2 pairs	all 4 angles
Examples					

To classify the quadrilateral at the right, identify the following characteristics.

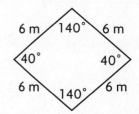

Number of parallel sides: 2 pairs

Number of congruent sides: all 4 sides

Number of congruent angles: 2 pairs

So, the figure is a rhombus.

Classify each quadrilateral. Write *quadrilateral, trapezoid, parallelogram, rectangle, rhombus,* or *square.* Answers may vary.

1.
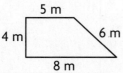

rectangle

2.
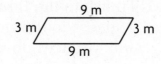

parallelogram

3.
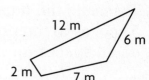

quadrilateral

4.
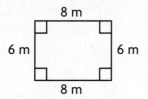

trapezoid

5.
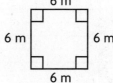

square

6.

rhombus

© Harcourt

Algebra: Transformations

When you move a figure, it is called a rigid transformation.
A translation is one type of transformation.

When you translate, or slide, a figure on a coordinate plane,
the coordinates change. The figure may move up or down, left
or right, or both. Here are three examples of translations.

3 spaces to the right

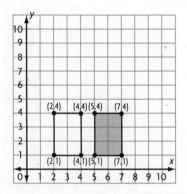

New ordered pairs:
(2,1) to (5,1),
(4,1) to (7,1),
(2,4) to (5,4),
(4,4) to (7,4)

2 spaces up

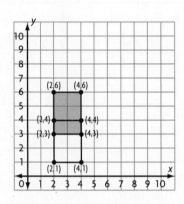

New ordered pairs:
(2,1) to (2,3),
(4,1) to (4,3),
(2,4) to (2,6),
(4,4) to (4,6)

**3 spaces to the right
and 2 spaces up**

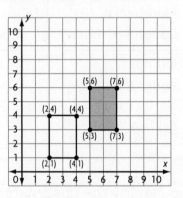

New ordered pairs:
(2,1) to (5,3),
(4,1) to (7,3),
(2,4) to (5,6),
(4,4) to (7,6)

Translate each figure. Draw the new figure with its coordinates.
Name the new ordered pairs. Check students' drawings.

1. Translate the figure
   5 spaces to the right
   and 4 spaces up.

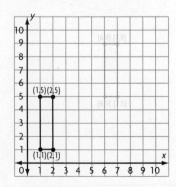

2. Translate the figure
   3 spaces to the right
   and 4 spaces down.

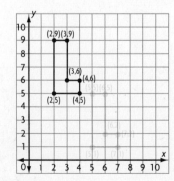

3. Translate the figure
   4 spaces to the left
   and 4 spaces down.

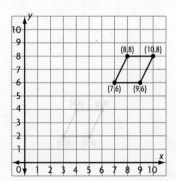

Solid Figures

A **prism** is a solid figure that has two congruent faces called **bases.** A prism is named by the polygons that form its bases.

The prism at the right is a hexagonal prism.

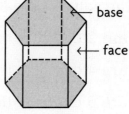

base

face

- The faces of this solid figure are rectangles.
- The bases of this solid figure are hexagons.

A **pyramid** is a solid figure with one base that is a polygon and three or more faces that are triangles with a common vertex. A pyramid is named by the polygon that forms its base.

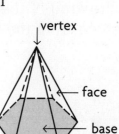

vertex

face

base

This is a hexagonal pyramid.

- The faces of this solid figure are triangles.
- The base of this solid figure is a hexagon.

Classify the solid figure. Then write the number of faces, vertices, and edges.

1.

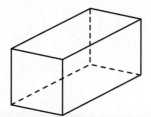

rectangular prism;

6; 8; 12

2.

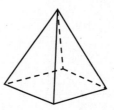

square pyramid;

5; 5; 8

3.

triangular pyramid;

4; 4; 6

Write the name of the solid figure.

4. I have a base with 5 equal sides. My faces are 5 triangles.

pentagonal

pyramid

5. All 6 of my faces are squares.

cube or

square prism

6. I have 2 congruent pentagons for bases. I have 5 rectangular faces.

pentagonal

prism

Draw Solid Figures from Different Views

A solid figure looks different when it is viewed from different positions.

Look at the solid figure at the right.

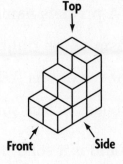

- There are 2 cubes in the top layer.

- There are 4 cubes in the middle layer.

- There are 6 cubes in the bottom layer.

This is a drawing of the figure viewed from the top.	This is a drawing of the figure viewed from the side.	This is a drawing of the figure viewed from the front.
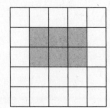	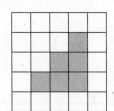	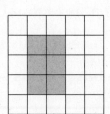

For 1–6, use the figure on the right.
Tell how many cubes are in each row.

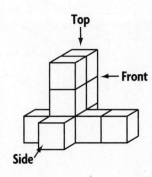

1. top layer   2 cubes

2. middle layer   2 cubes

3. bottom layer   6 cubes

Draw the figure from different views.

4. top view   **5.** side view   **6.** front view

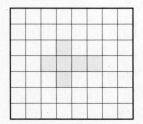 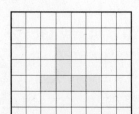 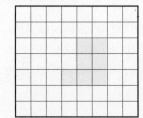

Problem Solving Skill
Make Generalizations

When you generalize, you make a statement that is true about a whole group of similar situations. Read the following problem.

Jesse is going camping. He will take with him: a box of cereal, paper towels, a flashlight, a soccer ball, and a tent shaped like a teepee. What polyhedrons will Jesse take on his camping trip?

1. Use what you know about each object to make a generalization. You can create a chart: Possible answers are given.

OBJECT	GENERALIZATION
cereal	Cereal usually comes in rectangular boxes.
paper towels	Paper towels come in a roll, which is a cylinder.
a flashlight	Most flashlights are cylinders.
a soccer ball	A soccer ball is a sphere.
tent	A teepee is shaped like a cone.

2. Sort the shapes.

prisms: box of cereal; sphere: soccer ball; cone: tent; cylinders: paper towels, flashlight.

3. Solve the problem. Which shapes are polyhedrons? Explain.

the box of cereal; its faces are all polygons.

4. Describe the strategy you used.

I made a table comparing the objects to polyhedrons.

Use what you know about each object to make a generalization. Then solve.

5. Annie takes a book, 2 cans of fruit juice, and a wedge pillow to the beach. What solid figures does she have? How many of these are polyhedrons?

1 rectangular prism (the book), 2 cylinders (the juice cans), and

1 triangular prism (the wedge pillow); two, the book and the pillow.

Customary Length

You can measure more precisely by using smaller units of measure.

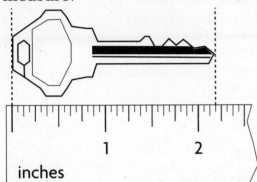

inches

Measured to the nearest inch: 2 in.

Measured to the nearest $\frac{1}{4}$ inch: $2\frac{1}{4}$ in.

Measured to the nearest $\frac{1}{16}$ inch: $2\frac{3}{16}$ in.

So, the measurement to the nearest $\frac{1}{16}$ inch is most precise.

For 1–5, use a customary ruler to measure your textbook. Answers will vary.

1. to the nearest inch: height _____   width _____

2. to the nearest $\frac{1}{2}$ inch: height _____   width _____

3. to the nearest $\frac{1}{4}$ inch: height _____   width _____

4. to the nearest $\frac{1}{8}$ inch: height _____   width _____

5. to the nearest $\frac{1}{16}$ inch: height _____   width _____

6. Which is the most precise measure? least precise measure?

to the nearest $\frac{1}{16}$ inch; to the nearest inch

_____

Measure each line segment to the nearest $\frac{1}{16}$ inch.

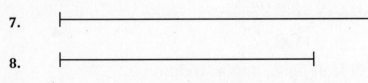

7. ⊢——————————————⊣ $3\frac{7}{16}$ in.
_____

8. ⊢————————————⊣ $2\frac{3}{4}$ in.
_____

9. ⊢——————————⊣ $1\frac{7}{8}$ in.
_____

10. ⊢———————————⊣ $2\frac{1}{2}$ in.
_____

11. ⊢———⊣ $\frac{11}{16}$ in.
_____

© Harcourt

Metric Length

Use your fingers to help you estimate metric length.

1 2 3 4 5 6 7 8 9 10

Compare the width of each of your fingers to 1 centimeter.
Is one of your fingers about 1 cm wide?

Use your fingers to help estimate the length of each object. Then
use a ruler to measure to the nearest centimeter and millimeter.
Estimates will vary.

1.

2.

5 cm; 5 cm; 47 mm   3 cm; 3 cm; 27 mm

3.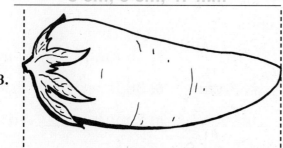

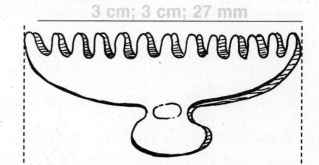

7 cm; 7 cm; 71 mm   8 cm; 8 cm; 76 mm

5.

6.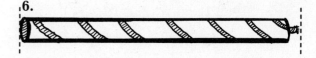

6 cm; 6 cm; 60 mm   8 cm; 8 cm; 77 mm

Use a ruler to draw a line segment of the given length. Check
students' lines.

7. 6 cm 3 mm   8. 3 cm 8 mm

9. 9 cm   10. 52 mm

Change Linear Units

Use a mental image to help you decide whether to multiply or
divide when changing linear units.

6 yd = ■ ft

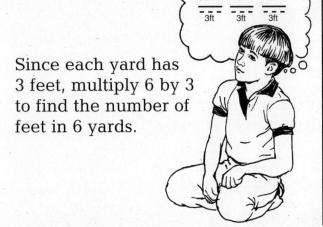

Since each yard has
3 feet, multiply 6 by 3
to find the number of
feet in 6 yards.

48 in. = ■ ft

Since each foot
has 12 inches,
divide 48 by 12 to
find the number
of feet in 48 inches.

Use a mental image to help you change the units.

1. 3 ft = ___36___ in.

2. 12 ft = ___144___ in.

3. 15 km = ___1,500___ m

4. 36 ft = ___12___ yd

5. 80 mm = ___8___ cm

6. 36 ft = ___432___ in.

7. 30 yd = ___90___ ft

8. 7 ft = ___84___ in.

9. 2 mi = ___3,520___ yd

Complete.

10. 3 ft = 2 ft □ in.

   3 ft = 2 ft + ___1___ ft

       = 2 ft + ___12___ in.

11. 3 km 9 m = 2 km □ m

   3 km 9 m = 2 km + ___1___ km + 9 m

           = 2 km + ___1,000___ m + 9 m

           = 2 km + ___1,009___ m

12. 7 cm 8 mm = 6 cm □ mm

_____18_____

13. 8 mi 30 yd = 7 mi □ yd

_____1,790_____

Find the sum or difference.

14. 2 ft 3 in.
   +4 ft 10 in.
    7 ft 1 in.

15. 2 ft 1 in.
  −   9 in.
   1 ft 4 in.

16. 8 m 4 cm
  − 5 m 80 cm
  2 m 24 cm

17. 5 m 13 cm
  +1 m 5 cm
  6 m 18 cm

Customary Capacity and Weight

You can change units of weight with multiplication or division.

Change larger units to smaller units by using multiplication.

$$3 \text{ lb} = \blacksquare \text{ oz}$$

Pounds are larger than ounces, so multiply.

$$3 \times 16 = 48$$
$$\uparrow$$
(16 oz in 1 lb)

So, 3 lb = 48 oz.

Change smaller units to larger units by using division.

$$48 \text{ oz} = \blacksquare \text{ lb}$$

Ounces are smaller than pounds, so divide.

$$48 \div 16 = 3$$
$$\uparrow$$
(16 oz in 1 lb)

So, 48 oz = 3 lb.

Write *multiply* or *divide*.

1. When I change pounds to tons,

   I _____ divide _____.

2. When I change ounces to pounds,

   I _____ divide _____.

Customary Units for Measuring Weight
16 ounces (oz) = 1 pound (lb)
2,000 pounds = 1 ton (T)

3. When I change tons to pounds,

   I _____ multiply _____.

4. When I change pounds to ounces,

   I _____ multiply _____.

Multiply to solve.

5. 6 lb = _____ 96 _____ oz

6. 15 lb = _____ 240 _____ oz

7. 4 T = _____ 8,000 _____ lb

8. 1 T = _____ 32,000 _____ oz

Divide to solve.

9. 20,000 lb = _____ 10 _____ T

10. 128 oz = _____ 8 _____ lb

11. 80 oz = _____ 5 _____ lb

12. 14,000 lb = _____ 7 _____ T

Multiply or divide to solve.

13. 96 oz = _____ 6 _____ lb

14. 20 lb = _____ 320 _____ oz

15. 5 T = _____ 10,000 _____ lb

16. 12,000 lb = _____ 6 _____ T

Metric Capacity and Mass

Use the conversion table to determine whether
to multiply or divide to change metric units.

Change the unit.

5 liters = ■ metric cups

Think:

1 liter = **4** metric cups

Multiply by 4 to
change liters to
metric cups.

Metric Units of Capacity and Mass
1,000 mL = 1 L
250 mL = 1 metric cup
4 metric cups = 1 L
1,000 liters = 1 kL
1,000 mg = 1 g
1,000 g = 1 kg

So, 5 liters = 20 metric cups.

Change the unit.

1. 750 mL= ■ metric cups

250 mL= ___1___ metric cups

___divide___ by ___250___ to
change mL to metric cups

750 mL= ___3___ metric cups

2. 8.5 L= ■ mL

1 L= ___1,000___ mL

___multiply___ by ___1,000___ to
change L to mL

8.5 L= ___8,500___ mL

3. 5,000 g= ___5___ kg

4. 3 kL= ___3,000___ L

5. 7 L= ___28___ metric cups

6. 3,000 mg= ___3___ g

Time

You can calculate elapsed time by using a clock.

Think of a clock as a circular number line. Count the hours by ones and then count the minutes by fives.

Mark worked from 9 A.M. to 5:15 P.M. How many hours did Mark work?

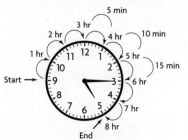

Count the hours and then count the minutes.

So, Mark worked 8 hours 15 minutes.

Use the clocks to determine the missing information.

1.

Begin
A.M.

End
P.M.

Elapsed time: _9 hr 30 min_

2.

Begin
P.M.

End
A.M.

Elapsed time: 16 hr 10 min

3.

Begin
A.M.

End
P.M.

Elapsed time: 9 hr 45 min

4.

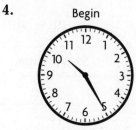

Begin
P.M.

End
A.M.

Elapsed time: _6 hr 25 min_

Problem Solving Strategy

Make a Table

You can make a table to show elapsed time.

Tara and Erin need to catch a school bus at 7:45 A.M. Before
they catch their bus, they need 15 minutes to shower, 15 minutes
to dress, 20 minutes to eat, and 10 minutes to walk to the bus.
For what time should Tara and Erin set their alarm?

A table can help organize the information. Work backward
from the final time. The starting time of one activity becomes
the ending time of the previous activity.

Activity	Start Time	End Time	Elapsed Time
Shower	6:45 A.M.	7:00	15 min
Dress	7:00	7:15	15 min
Eat	7:15	7:35	20 min
Walk to bus	7:35	7:45 A.M.	10 min

So, Tara and Erin must set their alarm for 6:45 A.M.

Make a table to solve.

Chenoa is planning a hike. He will hike for 40 minutes, eat for
20 minutes, hike for 30 minutes, rest for 10 minutes, and hike
for 40 minutes. He wants to end his hike at 1:30 P.M.

At what time should Chenoa start his hike? _____ 11:10 A.M. _____

Activity	Start Time	End Time	Elapsed Time
Hike	11:10	11:50	40 min
Eat	11:50	12:10	20 min
Hike	12:10	12:40	30 min
Rest	12:40	12:50	10 min
Hike	12:50	1:30 P.M.	40 min

Name _____

Perimeter

Since opposite sides of a rectangle are equal, you can use a formula to find the perimeter of a rectangle.

10 yd

5 yd

Perimeter $(P) = (2 \times l) + (2 \times w)$
$P = (2 \times 10) + (2 \times 5)$
$P = 20 + 10$
$P = 30$

So, the perimeter of the rectangle is 30 yd.

Since the sides of a regular polygon are equal, you can use a formula to find the perimeter of a regular polygon.

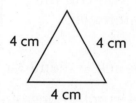

4 cm   4 cm

4 cm

Perimeter $(P) = $ (number of sides) $\times s$
$P = 3 \times 4$
$P = 12$

So, the perimeter of the triangle is 12 cm.

Find the perimeter of each polygon.

1.

4 cm

2 cm

length = __4 cm__ width = __2 cm__

$P = (2 \times $ __4__ $) + (2 \times $ __2__ $)$

$P = $ __8__ $ + $ __4__

$P = $ __12__ Perimeter is __12 cm__.

2.

7 ft

7 ft

side = __7 ft__

$P = $ __4__ $ \times $ __7__

$P = $ __28__ Perimeter is __28 ft__.

3.

3 m

side = __3 m__

$P = $ __5__ $ \times $ __3__

$P = $ __15__ Perimeter is __15 m__.

4.

3 in.

6 in.

length = __6 in.__

width = __3 in.__

$P = (2 \times $ __3__ $) + (2 \times $ __6__ $)$

$P = $ __6__ $ + $ __12__

$P = $ __18__ Perimeter is __18 in.__.

Reteach   RW141

Algebra: Circumference

The distance around a circular object is called its **circumference.**

A chord that passes through the center of a circle is a **diameter**.

If you know the diameter of a circle, you can find the circumference.

> **Remember. . .**
> The relationship of the diameter to the circumference of a circle, $C \div d$, is about 3.14 and is called *pi*.

Circumference ≈ diameter × 3.14, or $C \approx d \times 3.14$

Find the circumference of this circle.

≈ means "is approximately equal to".

Diameter = 4

Circumference ≈ diameter × 3.14

$\approx 4 \times 3.14$

≈ 12.56

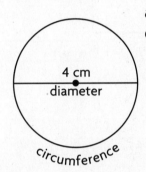

4 cm
diameter

circumference

The circumference is approximately equal to 12.56 cm.

The diameter of each circle is given. Multiply the diameter times 3.14 to find the circumference.

1.

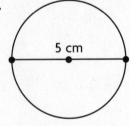

5 cm

__15.7 cm__

2.

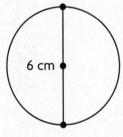

6 cm

__18.84 cm__

3.

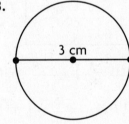

3 cm

__9.42 cm__

4.

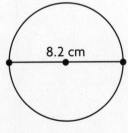

8.2 cm

__25.748 cm__

5.

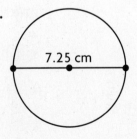

7.25 cm

__22.765 cm__

6.

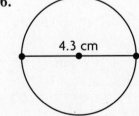

4.3 cm

__13.502 cm__

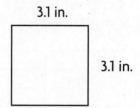
Algebra: Area of Squares and Rectangles

You can use a formula to find the area of a rectangle.

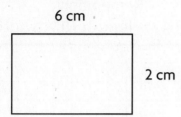

6 cm

2 cm

Area $(A) = l \times w$
$A = 6 \times 2$
$A = 12$

So, the area of the rectangle is 12 cm².

You can use a formula to find the area of a square.

3.1 in.

3.1 in.

Area $(A) = s \times s$
$A = 3.1 \times 3.1$
$A = 9.61$

So, the area of the square is 9.61 in.²

Find the area of these squares and rectangles.

1.

5 ft

5 ft

side = ____5 ft____

$A = $ ___5___ $\times$ ___5___

$A = $ ___25___

Area is ___25 ft²___.

2.

3 m

2 m

length $(l) = $ ____3 m____

width $(w) = $ ____2 m____

$A = $ ___3___ $\times$ ___2___

$A = $ ___6___

Area is ___6 m²___.

3.

2.5 yd

2 yd

length $(l) = $ ____2.5 yd____

width $(w) = $ ____2 yd____

$A = $ ___2.5___ $\times$ ___2___

$A = $ ___5___

Area is ___5 yd²___.

4.

8.4 m

8.4 m

side = ____8.4 m____

$A = $ ___8.4___ $\times$ ___8.4___

$A = $ ___70.56___

Area is ___70.56 m²___.

Relate Perimeter and Area

Rectangles with the same perimeter can have different areas.

Look at the rectangles below. Each rectangle has a perimeter of 24 cm, but their areas are different.

> **Remember. . .**
> Area (A) = length (l) × width (w)

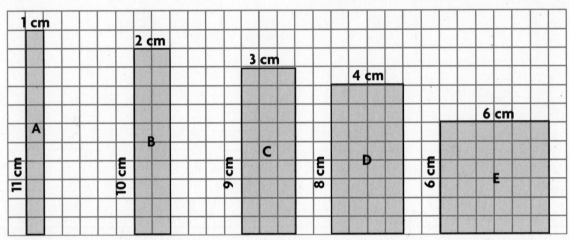

Rectangle A: Rectangle B: Rectangle C: Rectangle D: Rectangle E:
1 cm × 11 cm   2 cm × 10 cm   3 cm × 9 cm   4 cm × 8 cm   6 cm × 6 cm
Area = 11 cm²   Area = 20 cm²   Area = 27 cm²   Area = 32 cm²   Area = 36 cm²

Rectangle E is the rectangle with the greatest area, 36 cm².

Use the grid to draw rectangles for the given perimeter. Name Check students'
the length and width of the rectangle with the greatest area. drawings.

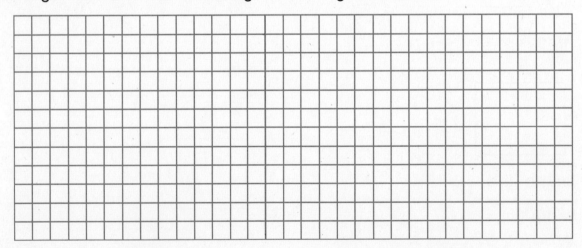

1. Perimeter = 12 cm

3 cm, 3 cm

2. Perimeter = 28 cm

7 cm, 7 cm

Algebra: Area of Triangles

Use what you know about the area of a rectangle to find the area of a triangle.

- Area of a rectangle equals length × width. ($A = l × w$)

- The area of a triangle is half the area of a rectangle with the same base and height. ($A = \frac{1}{2} × b × h$)

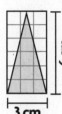

base (b) = 3 cm
height (h) = 6 cm

Area (A) = $\frac{1}{2} × b × h$

$A = \frac{1}{2} × 3 × 6$

$A = \frac{1}{2} × 18 = 9$

Area is 9 cm^2.

Find the area of these triangles.

1.

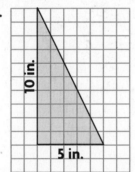

base (b) = ___5 in.___

height (h) = ___10 in.___

$A = \frac{1}{2} ×$ ___5___ × ___10___

$A =$ ___25___ Area is ___25 in.^{2}___.

2.

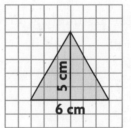

base (b) = ___6 cm___

height (h) = ___5 cm___

$A = \frac{1}{2} ×$ ___6___ × ___5___

$A =$ ___15___ Area is ___15 cm^2___.

3.

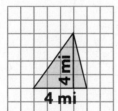

base (b) = ___4 mi___

height (h) = ___4 mi___

$A = \frac{1}{2} ×$ ___4___ × ___4___

$A =$ ___8___ Area is ___8 mi^2___.

4.

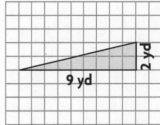

base (b) = ___9 yd___

height (h) = ___2 yd___

$A = \frac{1}{2} ×$ ___9___ × ___2___

$A =$ ___9___ Area is ___9 yd^2___.

© Harcourt

Algebra: Area of Parallelograms

Use what you know about the area of a rectangle to find the area of a parallelogram.

- Area of a rectangle equals length × width. ($A = l \times w$)

- The area of a parallelogram is equal to the area of a rectangle with the same base (length) and height (width). ($A = b \times h$)

You can use a formula to find the area of a parallelogram.

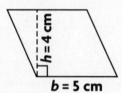

base (b) = 5 cm
height (h) = 4 cm

Area (A) = $b \times h$
$A = 5 \times 4$
$A = 20$
Area is 20 cm².

Find the area of these parallelograms.

1.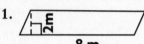
8 m

base (b) = _____8 m_____

height (h) = _____2 m_____

$A =$ ___8___ × ___2___

$A =$ ___16___ Area is ___16 m²___.

2.
5 yd

base (b) = _____5 yd_____

height (h) = _____6 yd_____

$A =$ ___5___ × ___6___

$A =$ ___30___ Area is ___30 yd²___.

3.
3 ft

base (b) = _____3 ft_____

height (h) = _____1 ft_____

$A =$ ___3___ × ___1___

$A =$ ___3___ Area is ___3 ft²___.

4.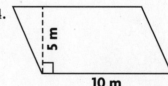
10 m

base (b) = _____10 m_____

height (h) = _____5 m_____

$A =$ ___10___ × ___5___

$A =$ ___50___ Area is ___50 m²___.

Name _____

Area of Irregular Figures

Count whole and half square units to find the area of an irregular figure on grid paper.

Count: 8 whole squares

12 half squares

Divide: 12 ÷ 2 = 6

Add: 8 + 6 = 14

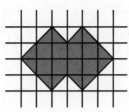

The area of the figure is 14 square units.

When an irregular figure on grid paper has partial squares that cannot be counted exactly, use the averaging method to estimate the area.

Count: 8 whole squares

8 partial squares

Divide: 8 ÷ 2 = 4

Add: 8 + 4 = 12

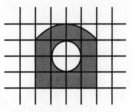

The area of the figure is about 12 square units.

Find the area. Each square is 1 cm².

1.

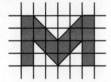

_____12_____ whole squares

_____8_____ half squares

_____12_____ ÷ 2 = _____6_____

_____12_____ + _____6_____ = _____18_____

Area = _____18_____ cm²

2.

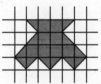

_____8_____ whole squares

_____12_____ half squares

_____12_____ ÷ 2 = _____6_____

_____12_____ + _____6_____ = _____18_____

Area = _____18_____ cm²

Estimate the area. Each square is 1 cm². Possible estimates are given.

3.

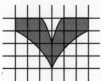

_____4_____ whole squares

_____14_____ partial squares

_____14_____ ÷ 2 = _____7_____

_____4_____ + _____7_____ = _____11_____

Area is about _____11_____ cm².

4.

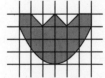

_____8_____ whole squares

_____14_____ partial squares

_____14_____ ÷ 2 = _____7_____

_____8_____ + _____7_____ = _____15_____

Area is about _____15_____ cm².

Problem Solving Strategy

Solve a Simpler Problem

Peter wants to paint a triangle with red paint on the playground. The height of the triangle will be 20 meters and the base 5 meters. Each container of red paint covers 10 square meters. How many containers of red paint will Peter need to paint his whole triangle?

Step 1

What does the problem ask? It asks how many containers of paint Peter will need.

Step 2

Find the area of the triangle.

Area $(A) = \frac{1}{2} \times$ base $(b) \times$ height (h)

$A = \frac{1}{2} \times 5 \times 20$

$A = 50$ The area is 50 m^2.

Step 3

Identify the number containers of red paint needed to cover 50 m^2.

Divide.

$50 \div 10 = 5$

So, Peter needs 5 containers of paint to paint the triangle on the playground.

Break these problems into simpler steps to solve.

1. Frank's house needs new carpet. The living room is 12 feet long and 13 feet wide. The dining room is 15 feet long and 11 feet wide. How many square feet of carpet will be needed?

   _____ 321 ft^2

2. Tom is laying new sod in his yard. His front yard is 20 yd by 15 yd, and his backyard is 20 yd by 20 yd. Sod is sold by the square foot. How many square feet of sod does Tom need?

   _____ 6,300 ft^2

Nets for Solid Figures

A **net** is a two-dimensional pattern for a
three-dimensional prism or pyramid.

Look at the net at the right.

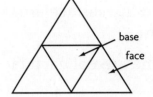

base
face

- It has 1 triangular base.

- It has 3 triangular faces.

Think about how you could
fold it to make a solid figure.

- It folds into a triangular pyramid.

Look at the second net at the right.

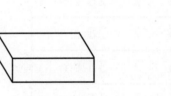
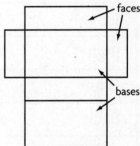

faces

bases

- It has 2 rectangular bases.

- It has 4 rectangular faces.

Think about how you could fold
it to make a solid figure.

- It folds into a rectangular prism.

Match each solid figure with its net.

1.

2.

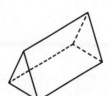

3.

_____ _____ _____

a.

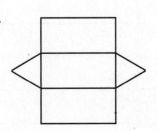

b.

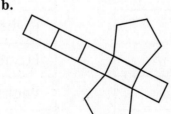

c.

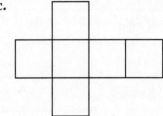

Surface Area

The surface area of a solid figure is the sum of the areas of its faces.

To find the surface area of a box, add the areas of the 6 faces.

Find the area of each face. Then add the areas.

Face	Area
Top	$5 \times 4 = 20$ cm^2
Bottom	$5 \times 4 = 20$ cm^2
Left	$4 \times 7 = 28$ cm^2
Right	$4 \times 7 = 28$ cm^2
Front	$5 \times 7 = 35$ cm^2
Back	$5 \times 7 = 35$ cm^2
Total Area	166 cm^2

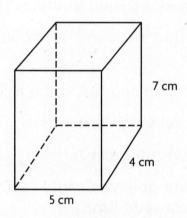

7 cm

4 cm

5 cm

Use the tables to find the surface area of each box.

1. 3 in. 5 in. 8 in.

2. 3 ft 2 ft 9 ft

Face	Area
Top	$5 \times 8 = 40$ in.^{2}
Bottom	$5 \times 8 = 40$ in.^{2}
Left	$3 \times 5 = 15$ in.^{2}
Right	$3 \times 5 = 15$ in.^{2}
Front	$3 \times 8 = 24$ in.^{2}
Back	$3 \times 8 = 24$ in.^{2}
Total Area	158 in.^{2}

Face	Area
Top	$9 \times 2 = 18$ ft^2
Bottom	$9 \times 2 = 18$ ft^2
Left	$2 \times 3 = 6$ ft^2
Right	$2 \times 3 = 6$ ft^2
Front	$9 \times 3 = 27$ ft^2
Back	$9 \times 3 = 27$ ft^2
Total Area	102 ft^2

Algebra: Volume

Volume is the amount of space a solid figure occupies or can hold. The formula for volume is:

$$\textbf{Volume} = \text{length} \times \text{width} \times \text{height}$$

Look at the rectangular prism at the right.

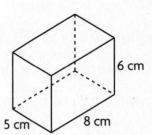

$V = l \times w \times h$

$V = (8 \times 5) \times 6$

$V = 40 \times 6 = 240$

So, the volume is 240 cm³.

To find a missing dimension, use the formula for volume.

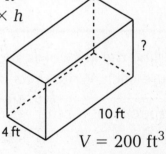

Step 1 Substitute the known values in the formula.

$V = l \times w \times h$

$200 = (10 \times 4) \times h$

Step 2 Multiply.

$200 = 40 \times h$

Step 3 Use Mental Math. Think: 40 times what number equals 200?

$200 = 40 \times 5$

$5 = h$

So, the height is 5 ft.

$V = 200$ ft³

Find the volume.

1. $l = 12$ yd, $w = 2$ yd, $h = 6$ yd

   $V = $ ___144 cu yd___

2. $l = 7$ m, $w = 12$ m, $h = 2$ m

   $V = $ ___168 cu m___

3. $l = 11$ cm, $w = 7$ cm, $h = 3$ cm

   $V = $ ___231 cu cm___

Find the missing dimension.

4. length = 6 in.

    width = 8 in.

    height = __5 in.__

    Volume = 240 in.³

5. length = __15 m__

    width = 5 m

    height = 2 m

    Volume = 150 m³

6. length = 10 ft

    width = __3 ft__

    height = 9 ft

    Volume = 270 ft³

7. length = 4 ft

    width = 5 ft

    height = __6 ft__

    Volume = 120 ft³

8. length = 6 cm

    width = __3 cm__

    height = 12 cm

    Volume = 216 cm³

9. length = __20 in.__

    width = 14 in.

    height = 7 in.

    Volume = 1,960 in.³

© Harcourt

Measure Perimeter, Area, and Volume

Keywords can help you decide upon the appropriate
unit of measure.

• Use *units* to measure the length of or distance around an object.

> Keywords: around, length, height, distance, perimeter

• Use *square units* to measure the area of an object.

> Keywords: cover, area, surface area

• Use *cubic units* to measure the volume of an object.

> Keywords: volume, capacity, fill, space

Substitute *inches, meters,* and so on for *units* when you are
given specific measurements.

Underline the keywords and tell the appropriate units to
measure each. Write *units*, *square units*, or *cubic units*.

1. capacity of a mug   **2.** paper to cover a box   **3.** length of a room

___cubic units___   ___square units___   ___units___

Underline the keyword and write the units you would use to
measure each.

4. surface area of this cube

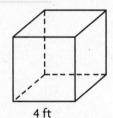

4 ft

_____ ft² _____

5. perimeter of this square

8 in.

_____ in. _____

6. volume of this prism

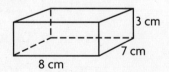

3 cm
7 cm
8 cm

_____ cm³ _____

7. area of this parallelogram

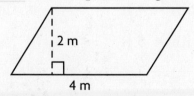

2 m
4 m

_____ m² _____

© Harcourt

Problem Solving Skill

Use A Formula

Paul wants to send some things to his brother at camp. He finds a box in his garage that is 12 inches long and 10 inches wide. It has a volume of 1,200 cubic inches. Paul wants to pack a container that is 14 inches high. Will the container fit in the box?

To answer the question, you need to find the height of the box.

Use the formula for volume. $V = l \times w \times h$

- Substitute. $1,200 = 12 \times 10 \times h$

- Multiply. $1,200 = 120 \times h$

- Use Mental Math. $1,200 = 120 \times 10$

The height is 10 inches. The answer is *no*, $h = 10$
the container will not fit in the box.

Use a formula and solve.

1. Rita wants to put a wallpaper border around her room. Her room is 11 ft by 13 ft. How many ft of border does she need to do the job? How many sq ft of carpet are required to carpet the room?

_____ 48 ft of border; _____

_____ 143 sq ft of carpet _____

2. Matthew needs to return a lamp that measures 15 inches by 12 inches by 8 inches. He has a box that is 16 inches long and 13 inches wide. It has a volume of 2,080 cubic inches. Is the box big enough for the lamp?

_____ Yes; the box is 10 in. high. _____

3. Mark needs to ship 500,000 cm³ cubic centimeters of peanuts. The shipping crate is 90 centimeters by 80 centimeters by 70 centimeters. Is it large enough to ship the peanuts? Explain.

_____ Yes; the volume is 504,000 _____

_____ cu cm. _____

4. How many square feet of carpet do you need to cover a 12-foot by 15-foot room? how many square yards?

_____ 180 sq ft; 20 sq yd _____

Understand Ratios

You can use decimal models to help find ratios. Ratios compare two quantities. There are three types of ratios.

Part to Whole

Shaded parts: 4

Total parts: 10

So, the ratio of part to whole is 4 to 10.

Whole to Part

Total parts: 10

Shaded parts: 6

So, the ratio of whole to part is 10 to 6.

Part to Part

Shaded parts: 3

Unshaded parts: 7

So, the ratio of part to part is 3 to 7 or 7 to 3.

Complete the ratios.

1.

shaded parts: _____8_____

total parts: _____10_____

part to whole ratio _____8 to 10_____

2.

shaded parts: _____6_____

unshaded parts: _____4_____

part to part ratio: _____6 to 4_____

3.

part to whole ratio: _____9 to 10_____

whole to part ratio: _____10 to 9_____

part to part ratio: _____9 to 1_____

4.

part to whole ratio: _____3 to 10_____

whole to part ratio: _____10 to 3_____

part to part ratio: _____3 to 7_____

Tell which type of ratio is expressed.

5.

5 to 5 _____ part to part _____

10 to 5 _____ whole to part _____

© Harcourt

Express Ratios

You can write ratios in three ways.

A **part to whole** ratio can be written:

  6 to 10   6:10 $\frac{6}{10}$

A **whole to part** ratio can be written:

  10 to 6   10:6 $\frac{10}{6}$

A **part to part** ratio can be written:

  6 to 4   6:4 $\frac{6}{4}$

Write each ratio in three ways. Then name the type of ratio.

1. 4 red counters to 3 green counters

  4 to 3; 4:3; $\frac{4}{3}$ part to part

2. 12 pencils to 6 pens

  12 to 6; 12:6; $\frac{12}{6}$ part to part

3. 2 soccer balls out of 11 balls

  2 to 11; 2:11; $\frac{2}{11}$ part to whole

4. 16 of 24 students are boys

  16 to 24; 16:24; $\frac{16}{24}$ part to whole

Write a, b, or c to show which ratio represents each comparison.

5. 3 red apples out of 8 apples

  a $\frac{3}{8}$ **b** 8:3   **c** 3 to 11

  a

6. 7 boys to 8 girls

  a $\frac{8}{7}$ **b** 7:8   **c** 7 to 15

  b

7. 8 baseballs to 13 basketballs

  a 21:8   **b** 13 to 8   **c** $\frac{8}{13}$

  c

8. 1 month out of 12 months

  a 1:11   **b** 12 to 1   **c** 1 to 12

  c

Write each ratio in two other ways.

9. 3:5   3 to 5; $\frac{3}{5}$

10. 11 to 13   11:13; $\frac{11}{13}$

11. 28 to 47   28:47; $\frac{28}{47}$

12. 14:6   14 to 6; $\frac{14}{6}$

13. $\frac{21}{4}$ 21 to 4; 21:4

14. $\frac{7}{19}$ 7 to 19; 7:19

Ratios and Proportions

You can use pictures to show equivalent ratios.

This shows the ratio 3:4.   This shows an equivalent ratio, 6:8.

You can also use fractions to show equivalent ratios.

$$\frac{3}{4} = \frac{6}{8} = \frac{9}{12} = \frac{12}{16} = \frac{15}{20} = \frac{18}{24}$$

Draw pictures to determine if the ratios are equivalent. Then
write *yes* or *no*. Check students' drawings.

1. 1:2 and 3:6 ___yes___   **2.** 3:4 and 4:6 ___no___

3. 2:3 and 3:5 ___no___   **4.** 2:5 and 4:10 ___yes___

Write two fractions that are equivalent to each ratio. Answers will vary.

5. $\frac{4}{5} =$ ___$\frac{8}{10}$___ $=$ ___$\frac{20}{25}$___   **6.** $\frac{9}{2} =$ ___$\frac{18}{4}$___ $=$ ___$\frac{27}{6}$___

7. $\frac{11}{12} =$ ___$\frac{22}{24}$___ $=$ ___$\frac{33}{36}$___   **8.** $\frac{6}{10} =$ ___$\frac{3}{5}$___ $=$ ___$\frac{60}{100}$___

9. $\frac{8}{6} =$ ___$\frac{4}{3}$___ $=$ ___$\frac{12}{9}$___   **10.** $\frac{3}{7} =$ ___$\frac{6}{14}$___ $=$ ___$\frac{15}{35}$___

Write two ratios that are equivalent to each ratio. Possible answers are given.

11. 2:3 ___6:9; 10:15___   **12.** 3 to 4 ___6 to 8; 12 to 16___

13. 8 to 12 ___2 to 3; 24 to 36___   **14.** 5:7 ___15:21; 50:70___

15. 11:9 ___22:18; 44:36___   **16.** 16 to 4 ___4 to 1; 32 to 8___

17. 1:6 ___2:12; 3:18___   **18.** 2 to 10 ___1 to 5; 4 to 20___

19. 7:11 ___14:22; 28:44___   **20.** 2:6 ___1:3; 4:12___

Scale Drawings

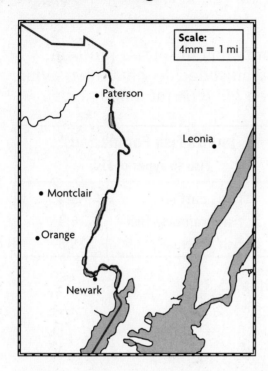

Scale:
4mm = 1 mi

• Paterson

Leonia

• Montclair

•Orange

Newark

You can use map scales and equivalent ratios to determine actual distances.

In this map of New Jersey, the scale is 4 mm = 1 mi. So, the ratio of millimeters to miles is 4:1.

The map distance from Paterson to Leonia is about 36 mm. What is the actual distance in miles?

Use equivalent ratios.

$$\frac{4}{1} = \frac{36}{n} \qquad \longleftarrow \quad 4 \times 9 = 36$$
$$\qquad\qquad \longleftarrow \quad 1 \times 9 = 9$$

Since $4 \times 9 = 36$, you would multiply 1×9.

So, the distance from Paterson to Leonia is 9 miles.

Use the map scale above and equivalent ratios to find the actual distance.

1. The map distance from Paterson to Newark is 48 mm.

   What is the distance in miles? _____12 mi_____

2. The map distance from Montclair to Orange is 12 mm.

   What is the distance in miles? _____3 mi_____

3. The map distance from Paterson to Orange is 42 mm.

   What is the distance in miles? _____10.5 mi_____

Use a map scale of 1 cm = 15 mi and equivalent ratios to complete the table.

4. Watertown to Belmont	6 cm	90 mi
5. Arlington to Bedford	4 cm	60 mi
6. Belmont to Avon	11 cm	165 mi
7. Franklin to Millis	3.5 cm	52.5 mi

© Harcourt

Problem Solving Skill

Too Much/Too Little Information

Sometimes you have *too much* or *too little* information to solve a problem. When you are given *too much* information, you must decide what information to use to solve the problem. When you are given *too little* information, you can't solve the problem.

Read the table carefully. Look at the question and decide if you have *too much* or *too little* information.

Jared's Fish Populations	
Fish to Types of Fish	
fish : catfish	5:1
fish : rainbowfish	25:3
fish : tetras	25:5

What is Jared's ratio of rainbowfish to tetras?

What information you **Know:**

• You know Jared's ratio of fish to rainbowfish is 25:3.

• You know Jared's ratio of fish to tetras is 25:5.

What information you **Don't Need:**

• You don't need the information on catfish.

You have too much information, so you can solve the problem. Jared's ratio of rainbowfish to tetras is 3:5.

Use the table to complete each problem.

1. How many red rainbowfish does Jared have for every one rainbowfish?

   ___can't solve this problem___

   What you Know:

   ___Jared's ratio of fish to rainbowfish is 25 to 3.___

   What you Don't Need:

   ___Jared's ratios of fish to catfish and tetras___

   What you Need to Know:

   ___information on Jared's red rainbowfish___

2. How many fish are there for every one tetra?

   ___5 fish for every one tetra___

   What you Know:

   ___Jared's ratios of fish to tetras is 25 to 5.___

   What you Don't Need:

   ___Jared's ratios of fish to catfish and rainbowfish___

   What you Need to Know:

   ___nothing___

Understand Percent

You can represent part of the whole by using percents.
Percent means "per hundred." 100 percent is the whole.

The 10 × 10 grid has 100 squares. Each square represents 1 percent.

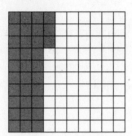

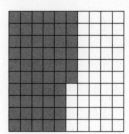

33 squares are shaded.
So, 33% of the squares are shaded.
67% of the squares are unshaded.

56 squares are shaded.
So, 56% of the squares are shaded.
44% of the squares are unshaded.

Write the percents for the shaded and unshaded squares.

1.
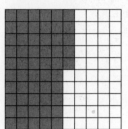

2.
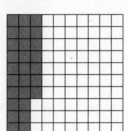

3.
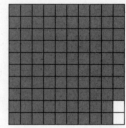

Percent shaded 55%

Percent unshaded 45%

Percent shaded 27%

Percent unshaded 73%

Percent shaded 98%

Percent unshaded 2%

Shade the 10 × 10 grid to show the percent. Check students' drawings.

4. 34%

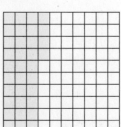

5. 69%

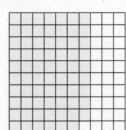

6. 82%

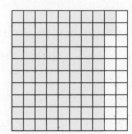

Relate Decimals and Percents

Percents and decimals both represent a part of a whole, or of 100.
You can use money to compare percents and decimals.

	Quarter	Dime	Nickel	Penny
Decimal:	$0.25	$0.10	$0.05	$0.01
Read:	twenty-five hundredths	ten hundredths	five hundredths	one hundredth
Ratio:	25 out of 100	10 out of 100	5 out of 100	1 out of 100
Percent:	25% of a dollar	10% of a dollar	5% of a dollar	1% of a dollar

Write a decimal and a percent to describe each total amount.

1. 1 quarter, 2 dimes

   decimal __$0.45__

   percent __45%__

2. 1 quarter, 1 dime, 1 penny

   decimal __$0.36__

   percent __36%__

3. 3 quarters, 3 pennies

   decimal __$0.78__

   percent __78%__

4. 8 dimes, 3 nickels, 2 pennies

   decimal __$0.97__

   percent __97%__

5. 12 nickels, 4 pennies

   decimal __$0.64__

   percent __64%__

6. 3 pennies

   decimal __$0.03__

   percent __3%__

Write the number as a decimal and as a percent.

7. forty-five hundredths

   _____0.45; 45%_____

8. twenty-one hundredths

   _____0.21; 21%_____

9. eighty-four hundredths

   _____0.84; 84%_____

10. seventy-two hundredths

   _____0.72; 72%_____

Relate Fractions, Decimals, and Percents

Percents can be written as decimals, or as fractions with 100 as the denominator.

- 45% means forty-five hundredths, or 0.45.

- 45% also means $\frac{45}{100}$.

To write a fraction in simplest form, divide the numerator and the denominator by the same number. Keep doing this until 1 is the only common factor.

- $45\% = \frac{45}{100} = \frac{45 \div 5}{100 \div 5} = \frac{9}{20}$

So, $45\% = 0.45 = \frac{9}{20}$.

To write a fraction as a percent, write a fraction with the percent as the numerator and 100 as the denominator.

- $\frac{37}{100} = 37\%$, or 0.37

- $\frac{3}{4} = \frac{3 \times 25}{4 \times 25} = \frac{75}{100} = 75\%$, or 0.75

Complete. Write each as a decimal, a percent, and a fraction in simplest form.

1. $0.35 = 35\% = \frac{35}{100} = \frac{35 \div 5}{100 \div 5} = \underline{\frac{7}{20}}$

2. $\underline{0.25} = 25\% = \frac{25}{100} = \frac{25 \div 25}{100 \div 25} = \underline{\frac{1}{4}}$

3. $\underline{0.20} = 20\% = \frac{20}{100} = \frac{20 \div 20}{100 \div 20} = \underline{\frac{1}{5}}$

4. $0.90 = \underline{90\%} = \frac{90}{100} = \frac{90 \div 10}{100 \div 10} = \underline{\frac{9}{10}}$

5. $0.16 = \underline{16\%} = \frac{16}{100} = \frac{16 \div 4}{100 \div 4} = \underline{\frac{4}{25}}$

6. $0.49 = 49\% = \frac{49}{100} = \frac{49 \div 1}{100 \div 1} = \underline{\frac{49}{100}}$

Complete. Write as a decimal and as a percent.

7. $\frac{1}{20} = \frac{5}{100} = \underline{0.05}$, or $\underline{5\%}$
        8. $\frac{3}{10} = \frac{30}{100} = \underline{0.30}$, or $\underline{30\%}$

9. $\frac{11}{100} = \frac{11}{100} = \underline{0.11}$, or $\underline{11\%}$
        10. $\frac{6}{25} = \frac{24}{100} = \underline{0.24}$, or $\underline{24\%}$

11. $\frac{2}{5} = \frac{40}{100} = \underline{0.40}$, or $\underline{40\%}$
        12. $\frac{3}{4} = \frac{75}{100} = \underline{0.75}$, or $\underline{75\%}$

Find a Percent of a Number

You can make a model to find a percent of a number.

Find 40% of 30.

Step 1	**Step 2**	**Step 3**
Use pieces of paper to represent 30.	Separate the pieces of paper into 10 equal groups. Each group represents 10%.	Separate 4 groups from the rest. These 4 groups represent 40%.

Since each group has 3 pieces of paper, 4 groups have 12 pieces of paper.

So, 40% of 30 equals 12.

You can find a percent of a number by changing the percent to a decimal and multiplying.

Step 1	**Step 2**
Change the percent to a decimal.	Multiply the number by the decimal.
40% = 0.40	0.40 × 30 = 12
	So, 40% of 30 equals 12.

Use a decimal to find the percent of the number.

1. 10% of 30

_____3_____

2. 20% of 50

_____10_____

3. 15% of 40

_____6_____

4. 20% of 60

_____12_____

5. 25% of 40

_____10_____

6. 3% of 18

_____0.54_____

Mental Math: Percent of a Number

You can use mental math to find a percent of a number.

Find 30% of 50 chips

Think: 30% = 10% + 10% + 10%

$$30\% \text{ of } 50 = (10\% \times 50) + (10\% \times 50) + (10\% \times 50)$$
$$\downarrow \qquad\qquad \downarrow \qquad\qquad \downarrow$$
$$(0.1 \times 50) + (0.1 \times 50) + (0.1 \times 50)$$
$$\downarrow \qquad\qquad \downarrow \qquad\qquad \downarrow$$
$$5 \quad + \quad 5 \quad + \quad 5 \quad = 15$$

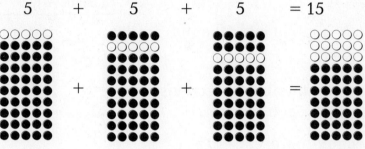

So, 30% of 50 = 15.

Use mental math to find the percent of each number.

1. 60% of 20

  12

2. 80% of 30

  24

3. 25% of 60

  15

4. 15% of 40

  6

5. 70% of 50

  35

6. 45% of 20

  9

7. 130% of 4,000

  5,200

8. 20% of 150

  30

9. 10% of 2,000

  200

10. 70% of 80

  56

11. 15% of 20

  3

12. 200% of 10,000

  20,000

13. 85% of 100

  85

14. 60% of 500

  300

15. 70% of 800

  560

Problem Solving Strategy

Make a Graph

You can make a graph to display percent data.

Step 1 Review your data. If your data show the relationship of parts to a whole, you can use a circle graph.

Step 2 Divide a circle into 10 equal sections.

Step 3 Label the number of sections that show each percent.

> 1 section represents 10%.
>
> 2 sections represent 20%.
>
> There are three 20% sections.
>
> 3 sections represent 30%.

Step 4 Label the percents and title the circle graph.

FAVORITE MAGAZINES

Magazine	Percent
Sports for Kids	20%
Around the World	30%
Media Talk	10%
Buy Smart	20%
Puzzle Power	20%

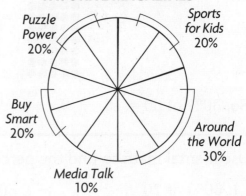

FAVORITE MAGAZINES

Use a 10-section circle and the data in the table to make a circle graph.
Check students' graphs.

1.

FAVORITE VACATIONS

Vacation Place	Percent
National Park	20%
Beach	20%
Amusement Park	40%
Foreign Country	10%
Famous City	10%

2.

FAVORITE HOBBIES

Hobby	Percent
Painting	20%
Collecting Stamps	10%
Making Models	30%
Collecting Stuffed Animals	30%
Other	10%

© Harcourt

Compare Data Sets

You can find a percent of a number to compare the results of two or more sets of data.

The circle graphs below show the results of surveys Joshua conducted.

Joshua's family	**Joshua's neighborhood**
There were 20 people surveyed.	There were 60 people surveyed.

FAVORITE ICE-CREAM FLAVORS IN JOSHUA'S FAMILY

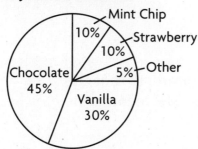

FAVORITE ICE-CREAM FLAVORS IN JOSHUA'S NEIGHBORHOOD

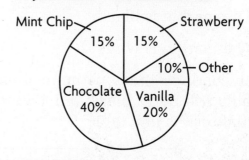

Joshua wants to know in which survey vanilla got more votes.

Joshua's family	**Joshua's neighborhood**
Find 30% of 20 people.	Find 20% of 60 people.

Joshua's family		**Joshua's neighborhood**	
Step 1	**Step 2**	**Step 1**	**Step 2**
Change the percent to a decimal.	Multiply the total number of people by the decimal.	Change the percent to a decimal.	Multiply the total number of people by the decimal.
$30\% = 0.30$	$0.30 \times 20 = 6$	$20\% = 0.20$	$0.20 \times 60 = 12$
So, 6 people voted for vanilla.		So, 12 people voted for vanilla.	

So, vanilla received more votes in Joshua's neighborhood survey.

For 1–2, use the circle graphs above.

1. In which survey did chocolate receive more votes? How many more votes?

    _____ neighborhood survey: _____

    _____ 15 more votes _____

2. If vanilla had received only 10% of the votes in Joshua's neighborhood, in which survey would vanilla have received more votes? Explain.

    _____ Neither; vanilla would have _____

    _____ received 6 votes in both surveys. _____

Probability Experiments

A box contains 6 black marbles and
2 white marbles.

Experiment: Shake the box, and pull a marble.
Record the color; then replace the marble.

There are two possible **events** for this
experiment.

- The marble is black.

- The marble is white.

Tom conducts the experiment 16 times. He predicts that the
events will be 12 black and 4 white marbles. He records the
actual results in a table.

MARBLE EXPERIMENT		
Events	**Black**	**White**
Predicted frequency	12	4
Actual frequency	⁄⁄⁄⁄ ⁄⁄⁄⁄ ⁄	⁄⁄⁄⁄

1. Is white or black a more likely event? Why?

   black; There are more black marbles than white marbles.

   _____

2. What were the actual frequencies after 16 trials?

   11 black marbles; 5 white marbles

   _____

3. Explain why Tom predicted 12 black and 4 white marbles.

   There are 3 times as many black as white marbles.

   _____

Outcomes

Your cafeteria offers a choice of tuna, turkey, or veggie sandwiches. You can also choose between white and wheat bread. What are your possible choices?

A **tree diagram** shows you all the possible choices.

Breads	Sandwiches	Choices
white	tuna	tuna sandwich on white bread
	turkey	turkey sandwich on white bread
	veggie	veggie sandwich on white bread
wheat	tuna	tuna sandwich on wheat bread
	turkey	turkey sandwich on wheat bread
	veggie	veggie sandwich on wheat bread

So, you have 6 possible choices.

---

1. Amanda must choose swimming a 25-meter, 50-meter, or 100-meter race. She can swim in either the freestyle or backstroke division. How many choices does Amanda have? Complete the tree diagram.

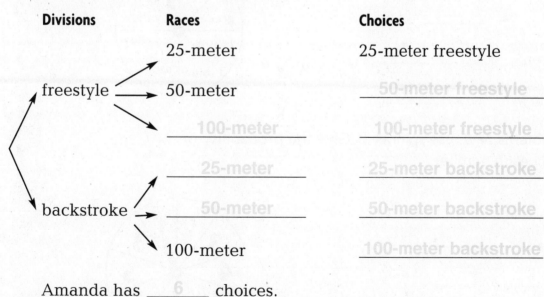

Divisions	Races	Choices
freestyle	25-meter	25-meter freestyle
	50-meter	50-meter freestyle
	100-meter	100-meter freestyle
backstroke	25-meter	25-meter backstroke
	50-meter	50-meter backstroke
	100-meter	100-meter backstroke

Amanda has ___6___ choices.

2. Brian's parents are buying a new car. They can choose a sedan or a minivan. Both cars are available in red or white. How many choices do they have? _4 choices_

Name _____

Probability Expressed as a Fraction

You can predict the **probability**, or chance, that an event will happen.

Ben has a spinner with six sections. The possible outcomes are spinning blue, spinning red, spinning yellow, or spinning green. What is the probability of spinning blue?

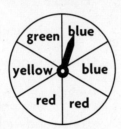

Probability of spinning blue = $\dfrac{\text{number of blue sections}}{\text{total number of sections}} = \dfrac{2}{6}$

So, the probability of spinning blue is $\dfrac{2}{6}$, or $\dfrac{1}{3}$.

For Problems 1–4, use spinner A. Give the probability of spinning each color.

1. blue ___$\dfrac{2}{8}$, or $\dfrac{1}{4}$___

2. red ___$\dfrac{2}{8}$, or $\dfrac{1}{4}$___

3. green ___$\dfrac{1}{8}$___

4. yellow ___$\dfrac{3}{8}$___

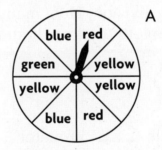

A

For Problems 5–8, use spinner B. Give the probability of spinning each number.

5. 1 ___$\dfrac{3}{10}$___

6. 2 ___$\dfrac{2}{10}$, or $\dfrac{1}{5}$___

7. 3 ___$\dfrac{2}{10}$, or $\dfrac{1}{5}$___

8. 4 ___$\dfrac{2}{10}$, or $\dfrac{1}{5}$___

9. 5 ___$\dfrac{1}{10}$___

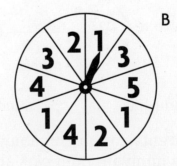

B

Compare Probabilities

You can compare the probabilities of events to determine whether one event is more likely than another.

Kim has a bag of marbles with 6 blue and 3 green marbles. Which color marble is she more likely to pull?

First, find the probability of each event.

Probability of blue $= \dfrac{\text{number of blue marbles}}{\text{total number of marbles}} = \dfrac{6}{9}$

Probability of green $= \dfrac{\text{number of green marbles}}{\text{total number of marbles}} = \dfrac{3}{9}$

Then, compare the probabilities.

$$\dfrac{6}{9} > \dfrac{3}{9}$$

So, Kim is more likely to pull a blue marble.

For Problems 1–2, use the spinner. Find the probability of each event. Decide which event is more likely.

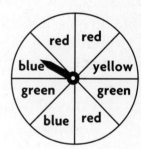

1. The pointer will land on blue; the pointer will land on red.

   Probability of blue $= \dfrac{\text{number of blue sections}}{\text{total number of sections}} = \dfrac{2}{8}$

   Probability of red $= \dfrac{\text{number of red sections}}{\text{total number of sections}} = \dfrac{3}{8}$

   More likely event: ____red____

2. The pointer will land on yellow; the pointer will land on green.

   Probability of __yellow__ $= \dfrac{\text{number of yellow sections}}{\text{total number of sections}} = \dfrac{1}{8}$

   Probability of __green__ $= \dfrac{\text{number of green sections}}{\text{total number of sections}} = \dfrac{2}{8}$

   More likely event: ____green____

Problem Solving Strategy

Make an Organized List

Making an organized list can help you determine the possible outcomes of a probability experiment.

Sharon has a coin and a spinner divided into two sections: red and yellow. She will toss the coin and spin the spinner. What are the possible outcomes? How many are there?

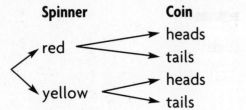

Spinner	Coin	Outcomes
red	→ heads	red and heads
	→ tails	red and tails
yellow	→ heads	yellow and heads
	→ tails	yellow and tails

So, there are 4 possible outcomes.

Make an organized list to solve.

1. Jereme is conducting a probability experiment with a coin and a bag of marbles. He has 3 marbles in the bag: 1 red, 1 purple, and 1 brown. He will replace the marble after each turn. How many possible outcomes are there for this experiment? What are they?

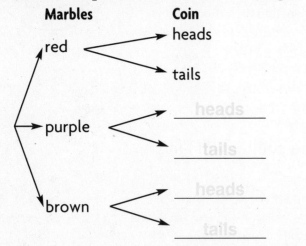

Marbles	Coin	Outcomes
red	heads	red and heads
	tails	red and tails
purple	heads	purple and heads
	tails	purple and tails
brown	heads	brown and heads
	tails	brown and tails

There are ___6___ possible outcomes.

2. Sarah has 10¢. How many different combinations of coins could she have? _____4 possible combinations_____